Ancient Indian Astronomy

ENIGMATIC, WONDERFUL & GLORIOUS TRADITION

NILESH NEELKANTH OAK
RUPA BHATY ✦ LEENA DAMLE

Translated by: Ashwini Sane

Sakāl Publications

Sakal Publications

PRACHIN BHARATIYA KHAGOLVIDNYAN

First Published in Marathi by Sakal Media Pvt. Ltd., India, 2025

ANCIENT INDIAN ASTRONOMY

Neelesh Oak, Rupa Bhaty, Leena Damle

Sakal Media Pvt. Ltd.
595, Budhwar Peth, Pune– 411002, India
www.sakalmediagroup.com
sakalpublications.com
sakalprakashan@esakal.com

First Edition: April, 2026

ISBN: 978-81-69355-47-6

Editor: Akshay Bhide
Cover Design & Layout: Yashodhan Lovalekar
Translator: Ashwini Sane

Printed in India by Sakal Media Pvt. Ltd.

To
The 'Modern' Researchers of
Ancient Indian Knowledge System

Index

Part I: The Antiquity Of Indian Knowledge And Culture

Chapter 1: 'Modern' Science In Ancient Indian Texts
9

Chapter 2: Ancient Indian Calendar
24

Chapter 3: Time Determination Of The Epic Mahabharata
35

Chapter 4: Astronomical Evidences In The Ramayana
43

Chapter 5: Where Exactly Was Ravan's Lanka?
49

Chapter 6: Surya Siddhanta
55

Chapter 7: Determination Of The Time Of
The Rigveda And The River Saraswati
60

Part II: Ancient Indian Astronomy : Observations, Proofs And Conclusions

Chapter 8: Detailed Astronomical Observations
By Raja (King) Bruhadrath
79

Chapter 9: Krutika, Ruksha And Aaraksha
83

Chapter 10: Rohini Shakata Bhed –
Crossing the Cart of Rohini
86

Chapter 11: The Four Dancers Of Jupiter
And The Seven Hands Of Saturn
93

Chapter 12: Agastya, Nataraj and Apasmar
98

Chapter 13: Shishumar Chakra (Wheel)
108

Chapter 14: Jyotirlinga In Central India And Sampat (Equinox)
112

Part III: The Knowledge Tradtion During The Medievial Ages

Chapter 15: Vishwarupadarshan
Or Was It 'The Universe' In Dnyaneshwari?
123

Chapter 16: Astronomy In Dasbodh
129

Details of References / **145**

Part I

THE ANTIQUITY OF INDIAN KNOWLEDGE AND CULTURE

1

'Modern' Science In Ancient Indian Texts

All living generations of Indians are deracinated when it comes to comprehending the antiquity of Indian civilization, and especially that of Indian astronomy. Fortunately, there have been notable torchbearers of Indian astronomy throughout history, and it is only due to their tapasya that we have access to the ancient knowledge of Indian astronomy. Limiting myself to the last 150 years, I would like to mention the likes of Lokmanya Tilak, S. B. Dixit, V. B. Ketkar, P. V. Vartak, P. V. Holay, Anil Narayanan, K. Chandra Hari, K. V. Sarma, Jeevan Rao, who kept the field of Indian astronomy vibrant through their contributions in translations, interpretations, re-interpretations, empirical validation, and documentation of historical works. The list is by no means exhaustive, and there are numerous other distinguished names not included above.

Let us summarize the mainstream academic view of the antiquity of Indian astronomy, which I consider fraudulent, unscientific, and illogical. It begins with vague, faulty, and handwaving statements such as, "Some of the earliest roots of Indian astronomy can be dated to the period of the Indus Valley

Civilization or earlier." It builds on this flawed premise and then worsens. For example: "Astronomy later developed as a discipline of the Vedangas, with the study of the Vedas dating to 1500 BCE or older." Naturally, they date the 'Vedanga Jyotisha' to 1400–1200 BCE or later. This arrogance and folly continue when they claim, "Indian astronomy was influenced by Greek astronomy beginning in the 4th century BCE." And what is their evidence? 'Yavanjataka' and 'Romaka Siddhanta' which these academics describe as "a Sanskrit translation of a Greek text disseminated from the 2nd century." What nonsense! Their assertions persist: "Indian astronomy flowered in the 5th century with Aryabhatta, who represented the pinnacle of Indian astronomical knowledge," while conveniently ignoring that the 'Aryabhatiya' contains astronomical and terrestrial evidence pointing to 26000 years prior to our times. Yet these agenda-driven scholars falter when they fail to present evidence of sophisticated astronomy from the West or the Middle East that approaches Indian astronomy until after the medieval period. How, then, do they escape this stark reality? By declaring, "Later, Indian astronomy significantly influenced Muslim astronomy, Chinese astronomy, European astronomy, and others." I close this reflection on the deracination of Indian astronomy by noting that one professor from an Ivy League university made it his life's mission to craft the narrative that Indian astronomy was shaped by Greek influence to such an extent that he used Indian texts to reconstruct what he termed "lost Greek astronomy."

To correct this modern academic rot concerning Indian astronomy is no easy task. Nor is it the objective of this book. Instead, this book seeks to offer a few carefully chosen pointers to help readers begin comprehending the deep antiquity of Indian astronomy not merely as pure science for its own sake, but as a pragmatic yet precise science and technology, and as a tool for the builders of prosperity, agriculture and navigation through accurate weather prediction and reliable global positioning systems (GPS), along with the broader development of civilization in science, mathematics, technology, medicine, architecture, music, and Itihasa.

The development of Indian astronomy did not occur in isolation. Astronomy is one of the six essential tools (Vedangas) required to comprehend the Vedas; and it is one of the three essential tools for documenting Itihasa. Astronomy serves as a calendar to the individual (jiva), guiding knowledge and action in the pursuit of desire and purpose. Thus, from the beginning, the Indian engagement with astronomy was holistic, forming an integral part of the philosophical, intellectual, religious, and economic development of Indian civilization.

Our attempt to explore the deep antiquity of Indian astronomy is constrained by the available evidence and by our ability to triangulate scattered astronomical references that point to remote epochs. One may begin with the interpretation of the star Sirius on the ecliptic (from the 'Rigveda'), which may point to a time 2,00,000 years ago; yet we must await additional validation from non-astronomical evidence before asserting such an interpretation with confidence. On the other hand, the star Swati (Arcturus) provides decisive evidence that ancient Indians tracked the movement of stars across the sky for at least 60,000 years and possibly up to 86,000 years in antiquity. Even more fascinating than this depth of time is the existence of a continuous tradition communicating the journey of star Swati through the epochs of the 'Ramayana', the 'Mahabharata', and the intervening periods of these Itihasa events. This continuity can be validated by references to star Swati as a foreigner and traveler in ancient texts such as the 'Taittiriya Samhita', the 'Taittiriya Brahmana', the 'Valmiki Ramayana', and the 'Apastamba Grihya Sutras'. Further validation arises from the latitude measurements of Nakshatras recorded in the 'Surya Siddhanta', which allow us to pinpoint the specific time (~8000 BCE) when the documented measurements of the Nakshatra latitudes were made. The point is that even with scant references scattered across these varied texts of Indian civilization, a portrait emerges of a continuous and vibrant culture one without known parallel elsewhere in the world.

The text of the 'Surya Siddhanta' conceals further surprises. For example, the stated value for the latitude of the star Canopus may indicate epochs of 38,000 BCE, 14,500 BCE, 12000 BCE, 8000 BCE, 5500 BCE, 3000 BCE, and the most recent ~580 CE. When the insights of the 'Surya Siddhanta' are combined with astronomical evidence and allusions from the 'Ramayana' and the 'Mahabharata', another compelling conclusion emerges: the Nakshatra system employed today is the same as that used in the 'Mahabharata', the 'Ramayana', and the Brahmana texts, and it very likely began around 28,500 BCE. This is not to suggest that the Nakshatra system originated 30,000 years ago; rather, that it existed in some earlier form and underwent a reshuffle or recalibration around that epoch. This remains a subject for future research. As I explore Indian astronomy alongside my co-researchers - Rupa Bhaty, Anil Narayanan, Jeevan Rao, Akshay Sant, Nikhil Bhalerao, Mahesh Deshpande et al. the scope of inquiry continues to expand.

Two additional domains that may have shaped the pragmatic, accurate, and precise character of Indian astronomy are agriculture and navigation. Evidence for harvesting, cooking, storage, and processing of agricultural products in the Indian subcontinent can be traced as far back as 45,000 years. Research suggesting that the original home of all domestic varieties of mice lies in India has been postulated as connected to India's status as one of the oldest if not the oldest homelands of agricultural development and its diffusion to other parts of the world. Evidence of sophisticated ports and navigation extends the antiquity of navigation in India to 20,000 years ago and beyond. Sustainable agriculture demands an accurate calendar, and long-range navigation requires not only an accurate and precise calendar but also reliable GPS. These two imperatives appear to have played a decisive role in shaping Indian astronomy.

Astronomy was also essential for recognizing and aligning with changing weather patterns and seasons, associated diseases, health conditions, and epidemics - thus contributing to the development of medical sciences. Ayurveda may therefore be both a cause and an effect of the further refinement, sophistication, and accuracy of Indian astronomy. Astronomy informed timekeeping, rituals, seasonal food selection, and astrology.

This book is a collection of essays transcribed from the recorded lectures of the authors. At this stage, many readers may find themselves content with a deep appreciation for the practical, pragmatic, yet precise nature of Indian astronomy and its antiquity. If so, the book has achieved its objective. However, the authors also hope that many readers will regard this as merely the first step in their journey toward comprehending Indian astronomy, and will undertake the necessary swadhyaya and abhyasa to pursue this compelling discipline as a foundation to understand the development of what we call "modern science" today. Each essay in this book highlights a facet of Indian astronomy, and together they point to a tradition extending 40,000+ years and beyond. The objective is to inspire the reader to continue exploring this subject far beyond the pages of this book.

◆◆◆

This volume brings together essays that undertake a detailed examination of two interrelated subjects: the depth of ancient Indian knowledge, especially in astronomy and the remote antiquity of Indian history itself. India's civilizational memory, extending across many millennia, appears to have been obscured during the recent centuries of foreign invasions, as though deliberately effaced from collective consciousness. To recover that memory, and to present the historical record in its proper form for future generations, is therefore not merely desirable but necessary.

One of the principal arguments advanced by those who deny the extreme antiquity of Indian history concerns the supposed chronology of the beginning of agriculture. Their general assumption is that agriculture began approximately ten thousand years ago. According to this view, only after human beings became sedentary could villages, kingdoms, empires, prosperity, chariots, and other instruments of organized civilization have emerged. Agriculture, they argue, presupposes the abandonment of earlier nomadic existence and settlement in one place. Since the last Ice Age is held to have ended roughly 13,000 years ago, they further conclude that systematic cultivation before that period would have been impossible.

The factual situation, however, is otherwise. Between 12,800 and 11,600 years before the present occurred the climatic episode known as the Younger Dryas. If one examines the geographical distribution of its effects across the globe, it becomes evident that the Indian subcontinent was only minimally affected. There was no continental ice cover over the subcontinent; rivers continued to flow; the conditions necessary for organized life persisted. Consequently, the assumption that glacial conditions rendered agriculture impossible in India until that late date does not withstand scrutiny.

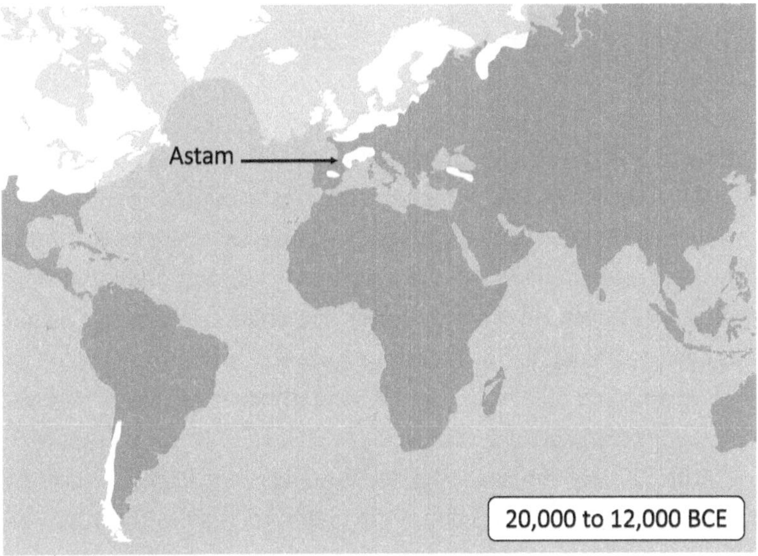

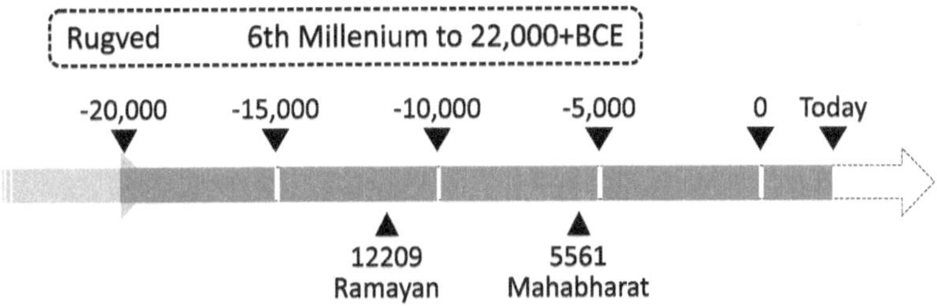

Timeline of Hindu Civilization

Both the Ramayana and the Mahabharata contain numerous references to astronomical events. Astronomical evidence, unlike conjectural historiography, cannot easily be dismissed. On the basis of such evidence, the antiquity of Indian history has already been argued with considerable force.

When one studies the records preserved in ancient Indian texts, one cannot but be struck by the range, subtlety, and depth of knowledge available to early Indian thinkers. A few illustrations may suffice.

Ancient Indians Knew... Theories of Cosmic Origin

Modern scientists have spent many decades investigating the profound question of how the universe originated. Today they command telescopes, artificial satellites, and computational systems; yet even now no definitive explanation of cosmic origin has been established. It is therefore remarkable that theories found in the Rigveda bear conceptual resemblance to what modern cosmology formulates under names such as the Big Bang theory.

The 129th hymn of the Tenth Mandala of the Rigveda—widely known as the Nasadiya Sukta, is devoted precisely to reflection on cosmic emergence.

Rigveda also preserves, through the fifteenth-century commentary of Sayana, a celebrated verse connected with the velocity of the sun:

तरणिर्विश्वदर्शतो ज्योतिष्कृदसि सूर्य।
विश्वमाभासि रोचनम्।।
योजनानां शते द्वे द्वे शते द्वे च योजने।
एकेन निमिषार्धेन क्रममाण नमोस्तुते।।४।।

(Rigveda Samhita 1.50.4)

Meaning:

O Sun, because your motion is swift, the distance to you is difficult to traverse.

Salutations to that Sun who crosses 2,202 yojanas in half a nimisha.

One day consists of 172,800 nimishas; thus one nimisha is taken as half a second.

One yojana is taken as 32,000 yards.

One yard equals 0.9144 metres.

Using these values, the resulting speed is 2.58×10^8 metres per second, a value strikingly close to the modern scientific estimate of the speed of light, namely 3×10^8 metres per second.

Ancient Indians Knew... The Stages of Biological Emergence

The historical evolution of the earth is also discussed in detail by Vasistha. Through the dialogue between Vasistha and the sage Kakabhushundi, regarded in tradition as one of the Chiranjeevis, the following descriptions appear:

O revered sage,

- I remember the earth devoid of rocks and trees, where neither grass nor creepers had yet arisen.
- I remember this earth submerged for eleven thousand years beneath heaps of ash.
- I remember a time when for four yugas the earth was covered with mountains and dense forests, without the movement of any living being.
- I have seen the earth covered for ten thousand years with mountain-like accumulations of the bones of dead daityas.

The possibility that such imagery corresponds, however loosely, to epochs after mass extinction, such as the end of the dinosaurs roughly 65 million years ago, invites reflection, even if not immediate literal acceptance.

Modern science has shown the earth to have passed through many radically different phases from its origin onward. The descriptions attributed to Kakabhushundi appear, step by step, to evoke precisely such successive transformations.

Ancient Indians Knew... Comets

Halley's Comet is now familiar because it returns roughly every seventy-six years. Ancient Indian observers too recorded cometary phenomena. More than that, they classified comets into three kinds:

1. Bhauma Ketu (terrestrial)
2. Antariksha Ketu (atmospheric)
3. Divya Ketu (celestial)

Varahamihira in the sixth century, and earlier Bhadrabahu, recorded observations concerning comets respectively in Brahma Samhita and Bhadrabahu Samhita. Later, Bhattotpala (937 CE), in the Brihat Samhita, listed numerous comets, assigning distinct names of sages to different appearances. Ancient Indian scholars also held that such comets returned periodically:

शतमेकाधीका मेके सहस्रमापरे वदन्ति केतूनाम।
बहुरूपामेकामेव प्राहमुनीर्नारदः केतुम्।।

(Brihat Samhitā 11.5)

Meaning:

Some, such as Parashurama, say there are 101 comets;

others, such as Garga, say there are 1,000;

while Narada says that there is but one comet appearing repeatedly in different forms and locations.

Ancient Indians Knew... The Rotation of the Earth

Aryabhata composed the Aryabhatiya at the age of twenty-three. In only 121 verses, the text presents extraordinarily dense astronomical insight.

Among its most significant statements are the following:

वृत्तभपञ्जरमध्ये कक्षापरिवेष्टितः खमध्यगः।
मृज्जलशिखिवायुमयो भगोलः सर्वतो वृत्तः।।

(Golapada, Aryabhatiya 4.6)

Meaning:

At the centre of circular celestial paths, suspended in space, is the terrestrial sphere—round on all sides, composed of earth, water, fire, and air.

That the earth is spherical is stated here with complete clarity—many centuries before such an idea became generally accepted in Europe.

Another verse reads:

अनुलोमगतिनौस्थः पश्यत्यचलं विलोमगं यद्वत्।
अचलानि भानि तद्वत् समपश्चिमगानि लङ्कायाम्।।

(Golapada, Aryabhatiya 4.9)

Meaning:

Just as a person moving forward in a boat sees stationary objects on the bank as moving backward, so too an observer at Lanka (that is, at the equator) perceives the fixed stars as moving westward.

Here Aryabhata rejects the older assumption that the earth is stationary and that the heavens rotate around it. Instead, he asserts that the earth rotates from west to east while the stars remain fixed—a formulation articulated nearly a thousand years before Nicolaus Copernicus.

Ancient Indians Knew... The Sphericity of the Earth and Gravitational Attraction

Bhaskara II writes in Siddhanta Shiromani:

आकृष्टि शक्तिस्तु महीयत् स्वस्थं गुरु स्वाभिमुखं स्वशक्त्या।
आकृष्यते तत्पततीव भाति समे समन्तात् क्व पतत्ययं खे।।

(Siddhanta Shiromani 19.6)

Meaning:

The earth, by its force of attraction, draws objects toward itself; thus they appear to fall. But if equal forces act from all sides in space, where would an object fall?

Five centuries later, Isaac Newton formulated universal gravitation. Newton is remembered universally; Bhaskaracharya, regrettably, far less so.

Ancient Indians Knew... The Earth as a Sphere Held by Attraction

The Yoga Vasistha contains striking cosmological reflection in the dialogue between Rama and Vasistha:

पिपीलिकानां महतां व्योम्नि वर्तुललोष्टके।
दशदिक्कमधः पादाः पृष्ठमूर्ध्वमुदाहृतम्।।१२।।

Meaning:

Imagine a great sphere suspended in space, with countless ants moving upon it. For each ant, the side of the feet is below and the back is above. Whichever way the sphere turns, the ants remain attached and do not fall.

Vasistha further clarifies that in the cosmos, the notions of "up" and "down" have no absolute meaning.

He continues:

प्रत्येकस्याण्डगोलस्य स्थितः कटकरत्नवत्।
भूताकृष्टिकरो भावः पार्थिवः स्वस्वभावतः।।३२।।

Meaning:

By the inherent attractive force of the terrestrial sphere, the elemental layers—earth, water, fire, air, and ether—remain bound to it.

This is scientific reasoning in remarkably direct form, preserved in an ancient philosophical text and too often neglected in modern reading.

Ancient Indians Knew... Eclipses

In many ancient cultures, solar and lunar eclipses were explained mythologically as celestial bodies being swallowed by demons. In India too, popular belief held that Rahu devoured the sun or moon during eclipse.

Yet Aryabhata explicitly states in the Aryabhatiya that a lunar eclipse occurs when the earth's shadow falls upon the moon, while a solar eclipse occurs when the moon comes between earth and sun.

Around 700 CE, the scholar Lalla of Dashapura also explained the precise causes of eclipses and rejected the popular Rahu explanation.

Ancient Indians Knew... Zero, Infinity, and Odd-Even Number Systems

The concept of zero is now universally acknowledged as one of India's great intellectual gifts to the world.

In the Yajurveda, odd numbers are enumerated in 18.19.24 and even numbers in 18.19.25:

ॐ एक च मे, त्रयश्च मे, पंच च मे...

and

द्वे च मे, चत्वारि च मे, षट्च मे...

These verses enumerate odd and even sequences not merely numerically but within ritual structure, indicating that number theory was integrated into sacrificial and philosophical contexts.

A. L. Basham writes in The Wonder That Was India that the concepts of zero and infinity were understood by ancient Indians with a clarity absent in much early Western thought.

The celebrated line of the Isha Upanishad—

पूर्णमदः पूर्णमिदं पूर्णात् पूर्णमुदच्यते।
पूर्णस्य पूर्णमादाय पूर्णमेवावशिष्यते॥

- is interpreted as expressing the idea that infinity, however divided, remains infinite.

Ancient Indians Knew... Trigonometric Principles

Trigonometry too is fundamentally rooted in Indian mathematical tradition. Texts such as the Surya Siddhanta, Mahabhaskariya, and Brahmasphuta Siddhanta preserve formulae and tabulations.

The Sulba Sutra states:

दीर्घ चतुरस्याक्ष्णया रज्जूः पार्श्वमानी तिर्यकमानी च यत्पृथग्भूते
कुरुतस्तदुभयं करोति।।

This states that in a right-angled triangle, the square on the diagonal equals the sum of the squares on the other two sides.

This is what later became known in Euclidean geometry as the theorem of Pythagoras. Yet, as Leopold von Schroeder notes in Pythagoras and the Indians, formulations equivalent to it appear in Indian sources centuries earlier.

How Was Such an Ancient History Preserved?

Three principal instruments made this preservation possible:

1. **Language** — a language that changed little across long durations: Sanskrit.
2. **Chronological Seal** — history requires a time stamp; astronomy provides the most rigorous such marker through planetary and stellar positions. Astronomy is time.
3. **Transmission Across Generations** — the guru–shishya oral tradition, supplemented by manuscripts and critical editions.

Because of these three, India's historical continuity survived.

Even today some ask whether the Ramayana and Mahabharata truly occurred. The question itself reveals the depth of historical deracination. Vast textual archives preserved by one's own civilization are doubted; assertions of invading powers are accepted without hesitation.

Had ancient Indian scholars not preserved innumerable astronomical references across texts, the antiquity of Indian civilization could not have been argued before the world with such force. India possesses a civilizational inheritance extending across many thousands of years. The Ramayana, the Mahabharata, and the Vedas—especially the Rigveda—contain multiple astronomical references through which the chronology of these Itihasa traditions may be examined.

How this is done will become clearer in the chapters that follow.

2

Ancient Indian Calendar

Whenever we look up at the night sky, we see innumerable stars scattered randomly across it. However, if we observe the sky day after day, month after month, we begin to realize that these stars are not randomly placed but arranged in a precise and consistent pattern, appearing in specific locations at specific times.

Since ancient times, humans observing these stellar patterns have named them after various animals or familiar objects on Earth. This made it easier to identify them while gazing at the night sky. Scientists across the world, while studying the heavens, use star charts that include 88 constellations officially recognized today.

To understand the antiquity of Indian culture, we must first comprehend the Indian system of calendar and time calculation. And to grasp this system, we need to briefly understand a few basic but important astronomical concepts:

Earth is called 'Bhugol'. 'Avkash' means space — the vast expanse containing planets and stars — and for astronomical purposes, it is sometimes treated as if bounded by a sphere (for example, when measuring distances between stars or determining the position of the Sun). In ancient astronomy, it was believed that the Earth lies at the center, surrounded by a vast sphere on which the stars appear fixed.

Earth is called 'Bhugol'.

The sphere around the Earth is 'Bhagol'.

The part of the sky visible to an observer on Earth is 'Khagol'.

The equator of the Earth, when extended outward onto this celestial sphere, forms the Celestial Equator (Vaishuvik Vrutta). The Earth's axis is tilted by about 23.5 degrees to this Celestial Equator, and the apparent path of the Sun across the sky is known as the Ecliptic (Ayanic Vrutta). The Ecliptic represents the path the Sun appears to follow as it moves through the constellations of the zodiac. It seems as though the Moon, the Sun, and the planets move from west to east relative to the stars along this Ecliptic.

The Earth exhibits three primary motions: the first is its rotation around its own axis (Parivalan); the second, its revolution around the Sun (Paribhraman); and the third, the slow wobbling of its tilted axis, known as precession (Paranchan). The Earth wobbles like a spinning top, and the rate of this axial movement is called 'Precession Speed' (Paranchan Gati).

The time required by the Earth to complete one revolution around the Sun is called a year. The precession of the equinoxes refers to the slow shift in the orientation of Earth's rotational axis — not its orbital revolution.

The Earth's equatorial plane forms an angle of about 23.5 degrees with the ecliptic plane. The two points where these circles intersect are called Equinoxes (Sampat Bindu). The Sun reaches these points twice each year — around March 20th or 21st, marking the Vernal Equinox (Vasant Sampat), and around September 22nd or 23rd, marking the Autumnal Equinox (Sharad Sampat). On these two days, day and night are of equal duration — 12 hours each.

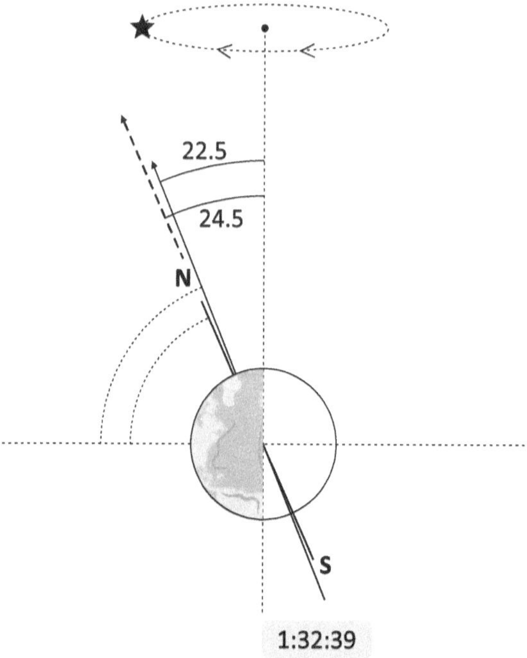

The Vernal Equinox is regarded as the starting point of the zodiac circle (Raashi Chakra). The zodiac sign Aries (Mesh) begins from this point. As the Vernal Equinox point shifts, the starting point progresses into the next zodiac sign.

It is not necessary that the Sun, during each Vernal Equinox, lies in the same specific constellation (Nakshatra). This variation occurs because of the Earth's Precession Speed. It has been observed that while the Sun moves eastward along the Ecliptic, the equinox shifts westward — in the opposite direction.

The time taken for this gradual change in equinoxes is approximately 26,000 years. If we divide 26,000 by 360, we get 72.22; therefore, the Earth's axis shifts by one degree every 72 years. Consequently, after every 72 years, the Vernal Equinox moves by one degree, and thus, the reference point of Uttarayan (the northward movement of the Sun) gradually shifts as well.

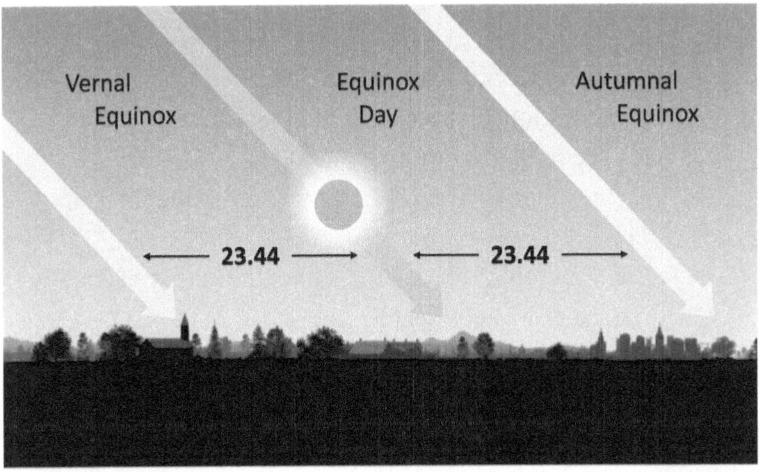

Vernal Equinox: Sun sets away from the West direction towards South
Autumnal Equinox: Sun sets away from the West direction towards North
Sun is exactly over the Equator on **Equinox day** and sets exactly at West.

Much like a spinning top that slowly changes the direction of its axis as it loses speed, the Earth's axis also shifts over time. Hence, the star aligned with this axis becomes the Pole Star for a certain period. The time taken for the axis to complete one full rotation is about 26,000 years. Due to this continuous motion, the Pole Star changes periodically.

The depth of thought and research devoted to understanding the concept of 'Time' in Hindu scriptures is unparalleled. Indeed, the term 'Kaal Chakra' — the Wheel of Time — embodies the belief that events recur cyclically as Time progresses. It is believed that this Wheel of Time consists of four Yugas: Satya, Treta, Dwapar, and Kali. Their durations are traditionally fixed as follows:

- The duration of Kali-Yug is 432,000 years.
- The duration of Dwapar-Yug is twice that of Kali-Yug, i.e., 864,000 years.
- The duration of Treta-Yug is three times that of Kali-Yug, i.e., 1,296,000 years.
- The duration of Satya-Yug is four times that of Kali-Yug, i.e., 1,728,000 years.

Together, these four Yugas constitute one Maha-Yug, totaling 4,320,000 years. It is believed that the evolution of life on Earth — from microbes to humans — required one Maha-Yug.

71 Maha-Yugs plus one Satya-Yug equals 3,08,448,000 years, which is known as one Manwantar. Our Sun requires one Manwantar to complete its revolution around the center of the Milky Way Galaxy (Akashganga).

14 Manwantar + 1 Satya-Yug = 4,32,00,00,000 years, which equals 1 Kalp (that is, 1000 Maha-Yug). This represents the current age of our universe. Such 2 Kalp make 1 day of Brahmadev, that is, 8.64 billion years. The universe is created during the day of Brahmadev and completely dissolved during his night. With the dawn of the next day, creation begins anew—and thus, the eternal wheel of time continues to turn.

This calculation pertains solely to astronomical cycles and observations; it is not connected to historical chronology.

Lunar and Solar Years

The Indian calendar system has, since ancient times, been a luni-solar one. While there are regional variations, we will focus here on its fundamental principles.

The system is based on the lunar day, or Tithi. A Tithi is defined as the time required for a 12-degree change in the angular distance between the Sun and the Moon. A Tithi may begin at any time of day, and its duration can vary between approximately 19 and 26 hours.

Each day, the Moon rises at different times and against different stellar backdrops. The same stars reappear behind the Moon roughly every 27 days. To map this cycle, ancient astronomers divided the 360° path of the Moon relative to the stars into 27 equal segments.

Each of these 27 divisions was given a distinct name—such as Ashwini, Bharani, Krutika, and Rohini—based on the most prominent or brightest star within that region. These segments are called Nakshatras (also known as Chandrapatni, from 'Chandra' meaning Moon and 'Patni' meaning wife).

In some instances, Nakshatras are named after individual stars, such as Swati and Chitra, while others are named after entire constellations, like Krutika, Pushya, Hasta, and Ashlesha.

Northward (Uttarayan) and Southward (Dakshinayan) Motion of the Sun

The Earth revolves around the Sun with its axis tilted at approximately 23.5°. Because of this tilt, at certain times the Northern Hemisphere leans toward the Sun, and six months later the Southern Hemisphere does. The hemisphere tilted toward the Sun experiences summer, while the opposite one experiences winter—this axial tilt being the fundamental cause of the seasons.

If we observe the sunrise and sunset daily, we notice that the Sun's position shifts gradually. When the Earth's North Pole is inclined toward the Sun, the sunrise moves northward. From around 21 December to around 21 June, the Sun's apparent motion is toward the north, a period known as Uttarayan. From around 21 June to around 21 December, the Sun appears to move southward—this period is called Dakshinayan.

27 NAKSHATRAS

1. Ashwini	10. Magha	19. Mula (Mul)
2. Bharani	11. Poorva Phalguni	20. Poorvashadha
3. Kruttika	12. Uttara Phalguni	21. Uttarashadha
4. Rohini	13. Hasta	22. Shravana
5. Mrugashirsha	14. Chitra	23. Dhanishtha
6. Ardra	15. Swati	24. Shatataraka
7. Punarvasu	16. Vishakha	25. Poorva Bhadrapada
8. Pushya	17. Anuradha	26. Uttara Bhadrapada
9. Ashlesha	18. Jyeshtha	27. Revati

There are specific constellations that lie along the path of the Sun, Moon, and planets, traced along the ecliptic. After a certain period, the Sun appears to move from one constellation to another. A total of twelve constellations lie in the Sun's path, and these constellations are known as "Raashi" (signs of the zodiac), each bearing a distinct name. These Raashis are named as follows:

1. Mesh (Aries)	5. Sinha (Leo)	9. Dhanu (Sagittarius)
2. Vrushabh (Taurus)	6. Kanya (Virgo)	10. Makar (Capricorn)
3. Mithun (Gemini)	7. Tula (Libra)	11. Kumbh (Aquarius)
4. Karka (Cancer)	8. Vrishchik (Scorpio)	12. Meen (Pisces)

This is the order of the zodiac signs. In a similar way, the Moon moves from one constellation to the next (Nakshatra). The constellation in which the Moon resides is also considered the Nakshatra of that day. One Raashi has two and one-quarter Nakshatras; two Raashis contain four and a half Nakshatras, eight Raashis have eighteen Nakshatras, and twelve Raashis together comprise twenty-seven Nakshatras. Each Raashi contains its own specific Nakshatras.

Among the Nakshatras, the first is 'Ashwini,' and among the Raashis, the first is 'Mesh.' Hence, the Ashwini Nakshatra and Mesh Raashi begin at the same point. Just as after Mesh Raashi come Vrushabh and Mithun in sequence, the Nakshatras too follow a fixed celestial order.

The Moon remains in each Nakshatra for about one day (roughly 24 hours), and the passage of that day is called a 'Tithi.'

In the Indian Lunisolar Calendar, the next unit is a Lunar Month (Chandramaas). Each lunar month consists of thirty Tithis, determined by the Moon's phases rather than fixed solar days.

1. Pratipada (1st day after New Moon / Full Moon)
2. Dwitiya (2nd day)
3. Tritiya (3rd day)
4. Chaturthi (4th day)
5. Panchami (5th day)
6. Shashthi (6th day)
7. Saptami (7th day)
8. Ashtami (8th day)
9. Navami (9th day)
10. Dashami (10th day)
11. Ekadashi (11th day)
12. Dwadashi (12th day)
13. Trayodashi (13th day)
14. Chaturdashi (14th day)
15. Purnima (Full Moon) in Shukla Paksha OR Amavasya (New Moon) in Krishna Paksha

These Tithis repeat themselves to complete a lunar month. They start as Pratipada, Dwitiya (First, Second) and so on and end with Amavasya (No moon night/ New Moon night).

Chandramaas Or Lunar Month

The duration of one lunar month is considered as either the period from one new moon to the next new moon (synodic month) or from one Poornima to the next Poornima. One lunar month is about 29.53 days.

The lunar month is named after the constellation in which the Moon resides during Poornima. Hence, from the name of the lunar month, we can immediately identify the corresponding constellation. For example, the Moon is near 'Chitra' during the Chaitra month, near 'Vishakha' during Vaishakh, and by observing the Nakshatra in which the full moon appears, one can determine the current Hindu lunar month.

Names Of Lunar Months

1. Chaitra	4. Ashadh	7. Ashwin	10. Paush
2. Vaishakh	5. Shravan	8. Kartik	11. Magh
3. Jyeshtha	6. Bhadrapad	9. Margashirsha	12. Phalgun

The Indian lunisolar calendar is 'Nirayana' (sidereal) based. One Nirayana year refers to the period during which the Sun, after completing one apparent revolution relative to the fixed stars, returns to the same position. One Nirayana year therefore consists of 12 solar months.

A lunar year, containing 12 lunar months, is shorter than the Solar Nirayana year. To reconcile the difference and keep the two years synchronized, an extra lunar month is periodically added to the lunar year.

The duration of a lunar month is about 29.5 days.

Hence, the duration of one lunar year = $12 \times 29.5 = 354$ days.

The duration of the solar year is approximately 365.2422 days (~365 days).

Thus, the difference between a solar and a lunar year is about $365 - 354 \approx 11$ days.

There is, therefore, a difference of about 10.88 days between the two. If this difference continues to accumulate, the lunar months would gradually drift away from the seasons. To correct this, whenever the difference amounts to roughly one lunar month (generally after 32–33 months), an additional lunar month known as 'Adhik Maas' is inserted.

This concept of an extra month kept the calendar aligned and precise. The practice of adding an intercalary month demonstrates the remarkable astronomical understanding and intellectual sophistication of our ancestors.

Although Tithis (lunar days) determine the calendar day and months are calculated according to the lunar cycle, the seasons are governed by the solar year. In an agrarian society, it is vital to maintain harmony between the lunar and solar systems, and our ancestors achieved this balance with great precision. Synchronization between lunar days and the solar year is essential for sowing and harvesting cycles.

For society to come together—to celebrate festivals, perform Yagyas (ritual fire ceremonies), and plan auspicious events such as marriages—it was imperative to have a correct and consistent calendar system. Thus, the ancient calendar served as a unifying framework for communal and agricultural life.

The Seasons Of The Lunar Month Change

The full moon appears in the constellation (Nakshatra) directly opposite the Sun. From that constellation, we can identify the lunar month. For instance, when the full moon is near the Chitra Nakshatra, the month is 'Chaitra'. When the Sun appears in one constellation, the full moon simultaneously appears in the opposite one.

In ancient Indian astronomy, various methods of time measurement were in use.

Shrimad Bhagwat Puran (Skanda 3, Adhyay 11, Shlokas 1–14) describes the following system of time measurement:

- 1 'Ahoratra' (Day and Night) = 8 Prahar = 24 hours
- 1 'Aha' (Night) = 4 Prahar = 12 hours
- 6 or 7 'Nadika Danda' = 1 Prahar = 3 hours
- 2 'Nadika Danda' = 1 'Muhurta' = 60 minutes
- 15 'Laghoo' = 1 'Nadika' = 30 minutes
- 15 'Kashtha' = 1 'Laghoo' = 2 minutes
- 5 'Kshan' = 1 'Kashtha' = 8 seconds
- 3 'Nimish' = 1 'Kshan' = 1.6 seconds
- 3 'Luv' = 1 'Nimish' = 0.53 seconds
- 3 'Vedh' = 1 'Luv' = 0.17 seconds
- 100 'Truti' = 1 'Vedh' = 0.56 seconds
- 3 'Trasarenu' = 1 'Truti' = 0.00056 seconds
- 3 'Anu' = 1 'Trasarenu' = 0.00019 seconds
- 2 'Paramanu' = 1 'Anu' = 0.000063 seconds
- 1 'Paramanu' = 0.000032 seconds

In this system of time measurement, 1 Muhurta equals 60 minutes. The term 'Hora' (hour) is derived from 'Ahoratra' by omitting the initial 'A' and the ending 'tra'. The division of time into 24 Horas aligns with astronomical principles and the seven-day week system.

Six or seven Nadikas constitute one Prahar, which means 24 Nadikas form one Aha (a complete day–night cycle). This flexibility in timekeeping may be attributed to seasonal variations, as the length of day and night fluctuates between 12 and 14 hours, depending on latitude and season.

3

Time Determination Of The Epic Mahabharata

In both the epics—the Mahabharata and the Ramayana, authored by Vyasa and Valmiki respectively, the sages have woven intricate word puzzles into their verses. Vaishampayana says, "Grantham granthi tada chakre munirgudham kutuhalat." This means that the epic Mahabharata is composed of numerous mysterious and adventurous stories, and within these stories lie many cleverly embedded riddles.

The Seven Sages (Saptarishi) And Arundhati

We are all familiar with Saptarishi (Ursa Major, a prominent asterism in the northern sky, commonly referred to as the Big Dipper). The Saptarishi forms a constellation with four stars that create a square and three stars extending from it like a tail. Among the three stars in the tail, the one in the middle is called Vasishtha (Mizar), and very close to it lies a faint companion star named Arundhati (Alcor).

The Saptarishi appears to move around the Pole Star (Polaris, the Dhruva Tara) in an anticlockwise direction due to Earth's rotation, with Vasishtha leading and Arundhati following behind.

However, Maharishi Vyasa, in the Mahabharata, describes a day prior to the war when he observes that "Arundhati is going ahead of Vasishtha."

अरुंधती तयाप्येष वसिष्ठः पृष्ठतः कृतः।

(Bhishma Parva, Adhyay 2, Shloka 31)

This Shloka means that Arundhati is carrying her husband behind her—that is, Vasishtha is following Arundhati.

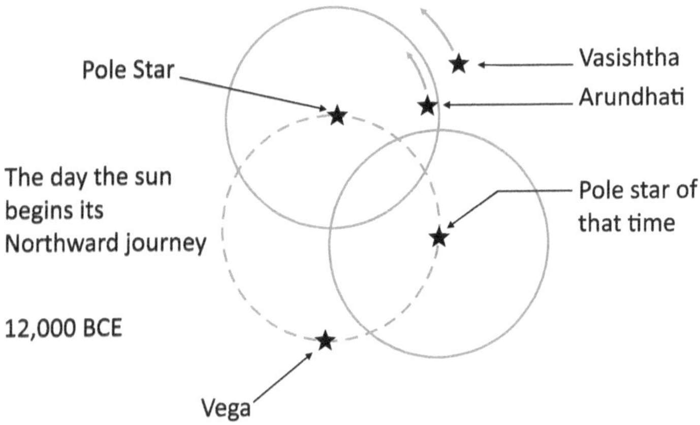

In the present age, and even in earlier recorded times, no such reversal of apparent position between Alcor and Mizar has ever been observed. When this intriguing astronomical puzzle is examined, it suggests that the Mahabharata period indeed dates back over 6,500 years, a conclusion drawn with considerable certainty.

According to the Indian Panchang (calendar system), the movements of the Sun and Moon, tithis (lunar days), nakshatras (lunar constellations), months, precession speed, and planetary positions all serve as valuable tools for recording historical events. This was effectively done by both Sage Valmiki and Sage Vyasa.

The term "Nakshatra" is defined as "Na Ksharati Iti Nakshatra," meaning "those that do not change their positions." However, ancient Indian sages and astronomers had long observed that even Nakshatras—which are fixed stars marking segments of the ecliptic—appear to shift their positions over time due to the precession of the equinoxes. They recorded these subtle positional changes, demonstrating the profound depth of their astronomical understanding and the great antiquity of Indian civilization.

To resolve the puzzle of the Mahabharata period, two major astronomical phenomena must be understood: "The Fall of Abhijit" (the former status of Vega as the Pole Star due to precession) and "The Epoch of Arundhati."

The Mahabharata contains around fifty descriptions of planetary positions—of observable planets such as Mars, Venus, Jupiter, Saturn, and Mercury. The position of Mars alone is recorded in eight different instances. Even if we analyze only the planetary motion of Mars, the same year consistently emerges—5561 BCE.

Mars is also known by several names: Angaraka, Dharaputra, and Lohitanga.

*मघा स्वंगारको वक्रः श्रवणेच बृहस्पतिः ।
कृत्वा चांगारको वक्र ज्येष्ठाया मधुसूदना ।
अनुराधा प्रार्थयते मैत्रं संशयमन्निव ।।
ध्रुवम् प्रज्वलितम् घोरं अपसव्य प्रवर्तते ।
चित्रा स्वात्यंतरेचैव धिष्ठितः परुषः ग्रह ।।
वक्रानुवक्र कृत्वश्च श्रवणे पावकः प्रभः ।
ब्रह्मराशी समावृत्त लोहितांगो व्यवस्थितः ।।

(Mahabharata, Bhishma Parva, Adhyay 3*)

Mars moves retrograde near Magha Nakshatra, and Jupiter does the same near Shravana Nakshatra, not once but multiple times. Mars also moves retrograde near Jyestha and Anuradha Nakshatras, then reverses its path between Swati and Chitra Nakshatras, doing so twice. Brahmarashi refers to the Abhijit Star (Tara)—which is not a permanent Pole Star, as different stars assume that role at different epochs due to Earth's axial precession. Abhijit was the Pole Star during the Ramayana, but by the time of the Mahabharata, it had become a constellation—and Mars was located within one Vrutta.

On the 18th day of the Mahabharata war, while Yudhishthira and Shalya were engaged in battle, it was noted that Yudhishthira observed three planets in the sky: Bhrigusuno (Venus), Dharaputra (Mars), and Shashijena (Mercury).

Understanding Retrograde Motion Of Planets

Let us understand what is meant by retrograde motion. Normally, planets in the solar system appear to move from west to east against the background of stars. Occasionally, however, they seem to move backward (from east to west). This apparent reversal is called retrograde motion.

The Sun, Moon, and planets move along the ecliptic plane. While the Sun and Moon follow regular orbital paths, their apparent speeds vary because Earth itself is in motion. At times, it seems as though they slow down, pause, and then reverse direction. The retrograde motion of planets is, therefore, an optical illusion—a result of our perspective from a moving Earth.

To illustrate, consider the planet Mars. Being an outer planet with a larger orbit than Earth's, Mars takes longer to complete one revolution around the Sun. Imagine Earth as a faster-moving vehicle overtaking a slower one—Mars. As Earth overtakes Mars, Mars appears to pause and then move westward, opposite its usual direction. In reality, the planets neither stop nor reverse their motion; they continue along their orbits at constant speed and direction. The phenomenon is merely an illusion caused by Earth's faster movement, making outer planets seem to move backward against the backdrop of the stars.

In our solar system, all planets, whether inner or outer, display an illusion of retrograde motion—that is, they appear to move in the opposite direction. However, the inner planets (Mercury and Venus) exhibit retrograde motion around their inferior conjunctions, while the outer planets (Mars, Jupiter, Saturn, Uranus, Neptune) do so when Earth overtakes them. All planets in our solar system, including the asteroids between Mars and Jupiter, exhibit retrograde motion when observed from Earth. In fact, nearly all asteroids trace retrograde loops as Earth overtakes them.

We have observed that the seasons in the lunar months shift over time. The Mahabharata war occurred during the Margashirsha month, and hence we celebrate Gita Jayanti—the auspicious day marking the Divine Revelation of the Bhagwat Gita—during this month. (In present times, Gita Jayanti falls in November-December.) Bhishma fell on the bed of arrows on the 10th day of the war. After that, he waited for approximately 95 days for the start of Uttarayan (around December 21) as per the Mahabharata. Adding 10 days to 95 gives 105 days; dividing this by 30 yields an approximate duration of three and a half lunar months.

Lunar months (e.g., Chaitra, Vaishakh, etc.) require roughly 2000 years to shift with respect to the seasons. For example, according to our current understanding, Chaitra and Vaishakh months correspond to Vasant Rutu (Spring). However, in reality, Vaishakh now falls in late spring or early summer, with nearly half of the winter season already past. Nearly 2000 years ago, both Chaitra and Vaishakh coincided with the Spring season, but due to precession, they now occur slightly later. For a one-month shift, approximately 2000 years are needed. Margashirsha now falls in November–December, near the winter solstice when Uttarayan begins. During the era of the Mahabharata, there was a four-month difference between Margashirsha and Uttarayan. Since a change of one month requires roughly 2000 years, a four-month shift corresponds to approximately 8000 years.

Thus, the Mahabharata war occurred around 8000 years ago.

Age Of Lord Krishna

Having deduced the date of the Mahabharata war, it is natural to inquire about Krishna. Krishna appears late in the Mahabharata, during Draupadi's Swayamvar (the ceremony of self-selecting a groom). At that time, both Krishna and Arjuna were 33 years old. Thirty-six years after the Mahabharata war, Krishna's Dwarka sank into the sea, marking the time of his demise, around 5525 BCE. The Mahabharata provides few details about Krishna, but the Harivansh and Bhagwat Puran offer extensive accounts. Harivansh recounts that a person named "Andhak" warned Kansa:

"There are ominous signs in both heaven and earth; he must improve his behavior, or his future will be grim."

What did Andhak foresee?

केतुना धूमकेतोस्तु नक्षत्राणि त्रयोदश: ।
भरण्यादीनि भिन्नानि नानुयांति निशाकरम् ।।

(Harivansh, Vishnu Parva, Adhyay 22)

Andhak observed a massive comet in the sky, spanning 13 nakshatras. Since there are 27 Nakshatras, this means the comet covered nearly half the celestial sphere—a rare astronomical phenomenon. Andhak noted: "The comet begins at Bharni Nakshatra and spans 13 more Nakshatras. It is so bright that its

22 March 5622 BCE

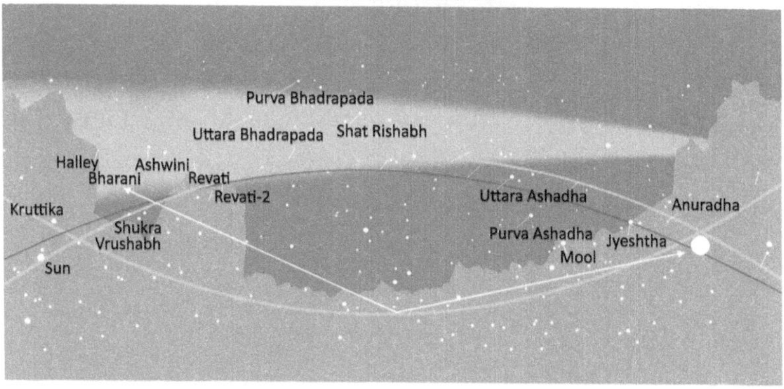

A huge comet occupying the area of 13 Nakshatras

luminescence obscures the moon's path through the Nakshatras." This description points to the year 5622 BCE in astronomical records.

At what age did Krishna kill Kansa? Nanadaraja had hidden Krishna in Gokul for approximately 11 years. Therefore, Krishna was 11 years old when he killed Kansa.

The comet was observed in the year 5622 + 11 = 5633 BCE, the year of Krishna's birth. Thus, Krishna's age at demise is: 5633 BCE – 5525 BCE = 108 years.

Modern science requires three elements to validate a claim, a process called triangulation: Explanation, Prediction, and Testing, applied within a theoretical and evidential framework. Bhagvan Patanjali refers to this in Sanskrit as:

प्रत्यक्षा अनुमाना आगम प्रमाणानि ।

Accordingly, to establish Krishna's birth year, we use astronomical knowledge, evidence from the Mahabharata, and accounts from Harivansh and Bhagwat Puran. When combined with Atma Pratiti (Self-Conviction), Guru Pratiti (Belief in the Teacher), and Shastra Pratiti (Belief in the Scriptures), the conclusion is validated. In Anugita, Krishna asserts: 'Tapobalmahin

Kevalam', meaning all accomplishments are attained through Tapa (Self-discipline or meditation) and Sadhana (dedicated practice).

The Mahabharatatatparyanirnaya, authored by Acharya Madhwa, records Krishna's age as 107 (saptamabdam shatottaram). Thus, for Krishna's age, all three criteria—Atma Pratiti, Guru Pratiti, and Shastra Pratiti—converge.

4

Astronomical Evidences In The Ramayana

Through the astronomical evidence presented in the epic Ramayana, it is possible to determine its time frame. Nilesh Oak states, "I was searching for a needle in a haystack spanning six thousand years, which could have enabled me to determine the time of the Ramayana, and by God's grace, I found that needle. This 'needle' was the utterance of Lakshman!"

नैर्ऋतं नैर्ऋतानां च नक्षत्रमभिपीड्यते ।
मूलम् मूलवता स्पृष्टं धूप्यते धूमकेतुना ।।

(Yuddhakand, Sarga 4)

Meaning:

"The demon's (Rakshas) root Nakshatra, which is 'Moola', whose deity is Nir-ruti, is in deep distress. A comet had appeared and is troubling the demon's root Nakshatra 'Moola.'"

Lakshman says this to Ram when he, along with Lakshman, Hanuman, and the vanarsena (an army of monkeys), starts toward Lanka.

I spent many days studying this 'directional needle' and, with the help of computers, created multiple astronomical charts for analysis. Through these charts, I was able to determine the period when that specific comet appeared over the Mula Nakshatra, which was in September, 12,209 years ago. Using this as a reference, I conducted further research. Hence, it is appropriate to call this comet the 'Lakshman Comet'.

Almost 575 astronomical events are described in the Ramayana. These observations can be divided into two categories. The first concerns events close to Earth, such as moon phases, planetary positions, lunar and solar eclipses, comets, and their effects. The second category involves effects arising from Earth's precession, such as the position of the sunset during Hemant Rutu, Sharad Rutu, and Chaitra Mahina, and other seasonal phenomena. At that time, the Pole Star was 'Abhijit' (Vega).

Let us examine other astronomical evidence in the Ramayana that helps determine its time frame. After finding Sita, Hanuman returns to Kishkindha, and everyone eagerly prepares to go to Lanka. Ram, Lakshman, and the entire vanarsena begin their journey. At night, they navigate by the stars, while Lakshman verifies their direction. He observes that the Pole Star, Brahmarishi, shines brightly. He describes it as follows:

ब्रह्मराशिर्विशुद्धश्च शुद्धाश्च परमर्षय: ।।
अर्चिष्मन्त प्रकाशन्ते ध्रुवं सर्वे प्रदक्षिणम् ।।

(Yuddhakand 4:48)

Meaning:

"Rama, many good omens are visible; we will surely win! Look at the Pole Star! It shines very brightly! The Saptarishi are revolving around it." (Note: The current Pole Star is not as bright. Today, the Saptarishi revolve around a different Pole Star.) The Pole Star around which the Saptarishi revolved was 'Abhijit' (Brahmarishi or Vega). Due to Earth's precession, calculations show that in 12,048 BCE, Abhijit was near the north celestial pole. Considering a margin of ±2,000 years, we conclude that the time when Abhijit served as the Pole Star was between 10,000 BCE and 14,000 BCE. Therefore, the Ramayana certainly could not have occurred after 10,000 BCE.

Some might question whether this is the only evidence for dating the Ramayana. To address this, let us examine additional observations. Ram, Lakshman, and Sita stayed for two years—from 12,211 BCE to 12,209 BCE—at Panchavati near Nasik. In the Valmiki Ramayana, Lakshman demonstrates keen observational skills. He records seasons, Nakshatras, sunrise and sunset positions, and the locations of various Nakshatras. He describes the sky as it appeared 14,000 years ago near present-day Nasik, Maharashtra, telling Ram and Sita: "Hemant Rutu is ongoing." Lakshman also describes everyday occurrences, seasons, and the locations of the sunset.

निवृत्ताकाशशयनः पुष्यनीता हिमारुणाः

(Aranyakand, Sarga 16, Shloka 12)

Meaning:

"The sunset during Hemant Rutu occurs over Pushya Nakshatra."

In the Kishkindha Kanda, when Hanuman leaps from Mahendra Parbat toward Lanka, he says:

भविष्यति हि मे पंथाः स्वातेः पंथा इवांबरे

(Kishkindha Kanda, Sarga 67, Shloka 19)

Meaning:

"I will not go straight but take a path like Swati Nakshatra." Hanuman observed this 14,000 years ago.

In the Ramayana, Lakshman tells Ram in Kishkindha (Hampi-Bellary, Karnataka):

"The Mula Nakshatra, situated in the southwest direction ('Nairutya'), is being troubled by a comet." From this, it can be deduced that the Ram–Ravan war occurred in 12,209 BCE.

In the Yajurveda's Taittiriya Brahmana, Swati Nakshatra is called 'Nishtayam', which means a traveler who has come from outside.

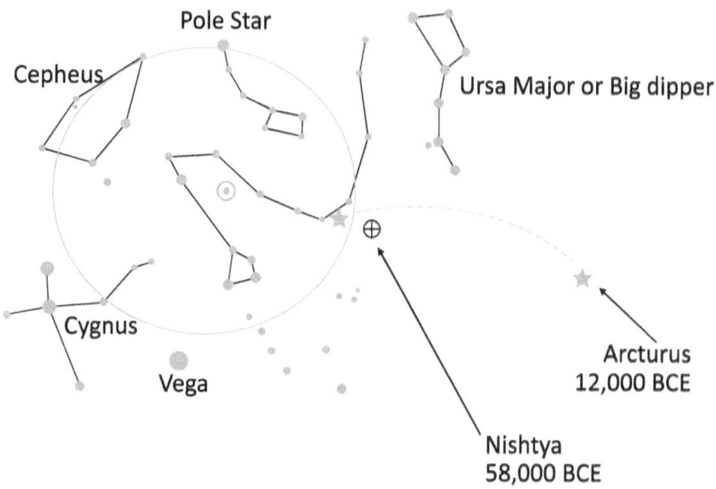

यां कामयेत दुहितरं प्रिया स्यादिति
तां निष्ट्याया दद्यात् प्रियैव भवति
नेव तु पुनरागच्छतीति ब्राह्मणवेक्षो विधिः

If we look at the sky 14,000 years ago, the star Swati was far to the north and served as the Pole Star. It travelled from the outer regions to become part of one of the 27 Nakshatras, as illustrated in the diagram. This exemplifies how much history is hidden in every word of the verse above.

It is stated in the 'Apasthamba Guhya Sutra': "Any father who believes that his daughter, once married, will go to her in-laws' house should love her husband and treat his home as her own, and she should not return. That father should marry off his daughter under the Swati Nakshatra." This is because a married girl moves from her family to her marital home, just like the star Swati.

Today, during the Hemant Rutu (November and December), after sunset, the following Nakshatras are visible: Mul, Poorvashadha, and Uttarashadha. For Lakshman to have seen the Pushya Nakshatra after sunset, we would have to go back 14,000 years. This implies that the events of the Ramayana occurred 14,000 years ago.

Ram's birth date, the 'Tithi', is Chaitra Shudha Navami. Currently, the lunar month of Chaitra falls in April. In the Ramayana, it is said, "It is a very auspicious month during which different fragrant and colorful flowers bloom." At present, this coincides with the onset of the monsoons when gardens are in full bloom and peacocks are active. Therefore, during the time of the Ramayana, the month of Chaitra must have followed the monsoons. Calculations based on Earth's precession indicate that this period corresponds to the start of the Hemant Rutu or the Sharad Rutu.

During the Ramayana, Agastya marked the Southern Celestial Pole, while the North Celestial Pole was Abhijit (based on the research of Shrimati Rupa Bhati). These two facts confirm that the Ramayana took place over 13,000 years ago. Multiple similar pieces of evidence in the Ramayana suggest that its events occurred around 12,000 BCE, with the Ram-Ravan war dating to 12,209 BCE.

The Pole Star shifts over time due to the Earth's precession. In the Mahabharata, numerous astronomical references exist, yet the Pole Star is never mentioned. This is because, during that period, no star occupied the position of the Earth's axis.

In contrast, the Ramayana beautifully describes a prominent Pole Star. Lakshman observes this star while moving from Kishkindha to Lanka. This star, Brahmarshi, corresponds to Abhijit or Vega. For Abhijit to have served as the Pole Star, according to Earth's precession, we must go back 14,000 years. Lakshman's observation thus aligns with the period of the Ramayana.

5

Where Exactly Was Ravan's Lanka?

Why ask, "Where is Ravan's Lanka?" Many assume that present-day Sri Lanka was Ravan's Lanka, and numerous sites in Sri Lanka are confidently associated with Ram-Ravan legends. So why does this question persist?

Numerous explorers have proposed various locations for Lanka, ranging from Sri Lanka, the equatorial Ling island, Sumatra, areas near Australia, and multiple regions in India: Assam, Maheshwara near the Narmada, Amarkantak in the Vindhya Range, towns near the Godavari Delta, Madhya Pradesh, Chhattisgarh, Gujarat, Sashti, Andhra Pradesh, and even Borivalli near Mumbai. The renowned historian and anthropologist Dr. H. D. Sankalia, then Head of the Department of Archaeological Survey of India, initially believed Lanka was near Chhattisgarh, Madhya Pradesh, though he later retracted his statement.

Dr. P. V. Vartak, in his book Vastav Ramayana, provides a logical explanation for these presumptions. Through scientific principles and systematic investigation, he demonstrates that the

Rakshas were called 'Niruti' because they lived in the southwest. The present 'Lanka', however, lies in the southeast direction (Agneya).

Let us examine Lanka's history. Ancient travelers, including Ptolemy and sailors of the Indian Ocean and Arabian Sea, knew it as the 'Sinhalese Island'. Arabs later called it 'Serendib'. Stone carvings indicate that Lord Buddha renamed it from "Sinhaldweep" to "Shri Lanka" around 2,500 years ago, and the British later called it 'Ceylon'. The name Sri Lanka was officially reinstated after 1972.

Approximately 2,000 years ago, Ptolemy mapped a place called "Taparoban," likely derived from the original name "Tamraparni." The topography on the Taparoban map closely resembles present-day Sri Lanka, particularly its mountain ranges. Some consider this island as Sumatra, while others regard it as a phantom island, as modern geologists have struggled to pinpoint its location near India. Ancient mapmaking was highly advanced, and these maps consistently show an island in the southeast direction—the location of modern Sri Lanka. Ptolemaic maps also depict Taparoban island below Sri Lanka, and these maps are probably over 2,000 years old.

Writers in the 8th and 12th centuries in India named Lanka and Sinhaldweep as two separate entities. Murari, a 9th-century writer, in his book Anargharaghav, writes, "Ram, while flying from Lanka, crosses Sinhaldweep after some time."

Rajshekhar, a 9th/10th-century writer of the drama Bal-Ramayana, writes that Ravan himself says, "'Lanka is to the south of Sinhaldweep,' and 'The king of it is different from Lanka.'" In the same drama, it is later written that when Ram is returning to Ayodhya by Pushpak Viman, he looks back and says, : "See, there is Lanka, the capital of the newly crowned King Bhibishan." These remarks occur after the Vimana has taken flight. Bhibishan, going further ahead, shows Sita the Sinwhal Mandal.

In addition, Rajshekhar's Kavayamimansa distinguishes Lanka and Sinhala as different places, with Lanka as the capital and Sinhala as the territory under Ravan's rule.

Of course, these are examples from written texts. Now, let us also examine the matter from a scientific perspective. Using geology, oceanography, and hydrology as the basis for study, we can view the subject from a different angle.

Fourteen thousand years ago, the sea level was lower than it is today, and thus Sri Lanka was connected by land to the Indian subcontinent. A map created by Somshekhar Ramaswamy from Tamil Nadu University shows that during the Last Glacial Maximum (20,000 years ago), the sea level was extremely low, exposing significant stretches of land beyond the current coastal region.

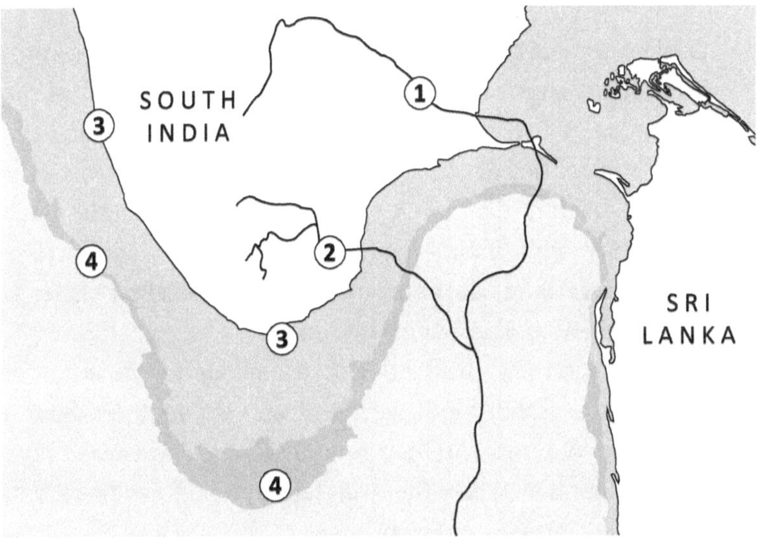

On the map, the area marked 3 represents the current Indian coastal region, while number 4 shows the coastal area at that time. Vast tracts of land were exposed, creating a continuous land bridge between India and Lanka. The number 2 marks the river Tamraparni, and 1 indicates the river Vaigayi.

When the ocean levels were down, Sri Lanka was connected to India by land

Lanka was thus connected to India when sea levels were low.

Now, let us consider the same subject during the time of the Mahabharata, before the Mahabharata war. When Dharmaraj (Yudhishthira) performed the Rajasuya Yagnya, he sent his four brothers in different directions. Sahadev, who went south, made a peace treaty with Bhibhishan. However, there is no precise information regarding Lanka's location. Notably, this Bhibhishan is not Ravan's actual brother but his descendant.

Sugreev describes the land and its layout to the team of monkeys sent to search for Sita. He instructs the southern-bound team: "When you cross the Malay mountain, you will reach near Mahendra Mountain, where the river Tamraparni flows; there, you will see the radiant Agastya (star)."

The arrow at the bottom of the map indicates the river Tamraparni, and its direction is significant. Today, this river no longer points toward Sri Lanka but toward the equator. Jatayu also advised Shri Ram to proceed southwest. After leaving Nashik (Maharashtra), Ram and Lakshman followed this southwest path and reached Kishkindha. If Ravan's Lanka were present-day Sri Lanka, their route would have passed through Nanded, Andhra Pradesh, and Tamil Nadu.

After receiving the Atmaling from Shivshankar, Ravan also traveled southwest from the Himalayas. Along the way, Ganesh took the Atmaling from Ravan and placed it near Gokarna, southwest of the Himalayas. The illuminated cave mentioned in the Valmiki Ramayana, where the monkey team got lost, is likely in the Thekkady region, where the terrain, though stunning, is extremely challenging.

Sugreev further instructs : "To reach Lanka, you must travel one hundred Yojanas below Mahendra Mountain." While the exact length of a Yojana is debatable due to historical variations, this indicates a vast distance over water. Today, no island exists at that location, though it is possible that the land was submerged, similar to Dwarka. Currently, one does not need to traverse a hundred Yojanas to reach Sri Lanka.

During the Ramayana period, the star Agastya was visible only in the southern part of India, as it was near the South Celestial Pole and served as a pole star.

According to ancient Indian astronomy, the zero-degree Rekha Vrutta (Prime Meridian) passed through Ujjain (modern-day reference for the Prime Meridian is Greenwich). Ujjain (Avanti), Kurukshetra, and Lanka were three key locations on this meridian, as repeatedly documented. Additionally, Lanka is described in the Surya Siddhanta as lying on zero degrees latitude and longitude, termed Niraksha (axis-less).

लंकावात्स्य्पुरावंतोस्थानेश्वरसुरालयान् ।
अवगाह्य स्थिता रेखा देशांतरविधायिनी ।।

(Suryasiddhanta)

In brief:

- Valmiki Ramayana indicates Lanka's location to the southwest.
- Ancient Indian astronomy places Lanka at zero degrees latitude and longitude (Niraksha).
- Ancient travelers and sailors documented Lanka and Sinhaldweep as distinct places.

Hence, researchers should revisit this study with a diverse perspective, and in time, the truth will emerge.

6

Surya Siddhanta

Surya Siddhanta is one of the most significant and profound works on ancient Indian astronomy, recognized as an epic worldwide. It comprises 14 chapters and 500 shlokas (verses). The language of the text is expansive, and to fully grasp the knowledge it contains, a reader must have a grounding in astronomy. An individual well-versed in the field will understand it more deeply.

सर्वत्रैव महीगोले स्वास्थानमुपरि स्थितम् ।
मन्यंते खे यतो गोलस्तस्य क्वोर्ध्वं क्व वाप्यधः ।।५३।।

(Bhugol Adhyay : 12)

Meaning:

"People living on the Earth unnecessarily feel that they are on top and that everything else is below. In reality, the Earth is a sphere and positioned in space. So, what is truly up or down?"

This indicates that our ancient scholars were aware that the Earth is spherical.

To locate a place on Earth, we use latitudes and longitudes. The Equator is defined as 0° latitude, the North Pole as +90°

latitude, and the South Pole as –90° latitude. Longitude is measured from the Prime Meridian at Greenwich (0°). On Earth's surface, latitudes and longitudes form an imaginary grid. Extending this grid outward, one can imagine it projected onto a celestial sphere. In modern astronomy, these coordinates correspond to declination (latitude equivalent) and right ascension (longitude equivalent), used to locate planets and stars in the sky.

Western scholars who studied the Surya Siddhanta have translated it into English. Ebenezer Burgess estimated that the text is nearly 1,500 years old, suggesting it was composed around 500 CE. The positions of various Nakshatras are clearly described. Calculating the celestial coordinates of these Nakshatras places the composition of Surya Siddhanta around 580 CE, though Wikipedia gives a broader timeframe of 800 BCE to 500 CE based on incomplete information.

The year of the Ram-Raavan war is given as 12,209 BCE. Let us now examine what the Surya Siddhanta says about this.

The original author of Surya Siddhanta is Mayasura. The knowledge imparted to him by the Sun's son or representative is called Surya Siddhanta.

मेरोरुभयतोमध्ये ध्रुव नभःस्थिते ।
निरक्षदेशसंस्थानामुभये क्षितिजाश्रिये ।।१२:४३।।
अतो नाक्षीच्छ्रयस्तासु धृवयोः क्षितिजास्थयोः ।
नवतिर्लंबकांशास्तु मेरावक्षांशकास्तथा ।।१२:४४।।

Meaning:

"There are two pole stars—one at the north celestial point and the other at the south celestial point. On the Earth's equator, these two stars appear on opposite sides."

The Surya Siddhanta's observation of two pole stars reveals a sophisticated understanding of celestial mechanics.

Earth revolves around the Sun in an elliptical orbit. At its closest point to the Sun, it reaches periapsis, and at its farthest, apoapsis. These points shift gradually over time.

The 12th chapter of Surya Siddhanta provides three proofs. At the time, two pole stars were observable, located oppositely in the northern and southern skies. Earth's axis is tilted by 23.5°, currently more precisely 23.4°, but this tilt oscillates between 22.1° and 24.5°.

Periapsis and Apoapsis: We are now familiar with Earth's precession. The axis points to a star designated as the pole star. Currently, Polaris serves this role in the north. The south has no prominent pole star today. However, Surya Siddhanta describes two pole stars. Around 2900 BCE, Shishumar (Thuban, Draco constellation) was the northern pole star. In the south, a bright star marked the celestial south pole. Over time, different sets of northern and southern pole stars appear. For instance, circa 12,000 BCE, Vega (Abhijit) was the northern pole star and Canopus (Agastya) the southern.

Which set of pole stars does Surya Siddhanta reference—2900 BCE or 12,000 BCE? The text mentions both timeframes.

Burgess dismissed ancient Indian astronomy, claiming that the slight difference between 23.4° or 23.5° and the 24° mentioned in Surya Siddhanta invalidates the text. He presumed the Surya Siddhanta was composed in 1,500 BCE.

भूमंडलात पंचदशे भागे दैवे तथाऽसुरे ।
उपरिष्ठाद व्रजत्यर्कः सौम्ययाम्यायनांतगः ।।१२:६८।।

Meaning: "It has been clearly stated in Surya Siddhanta that the Earth's axis was tilted by 24°."

Interestingly, the text specifies this tilt both around 12,000 BCE and near 2700 BCE.

Corroborating this, Surya Siddhanta also describes the seasons:

- Varsha (Monsoons): 21 June onwards (~July–August)
- Sharad (Autumn): Two months after Monsoons (~September–October)
- Hemant (Pre-Winter): (~November–December)
- Shishir (Winter): (~January–February)
- Vasant (Spring): (~March–April)
- Grishma (Summer): (~May–June)

अत्यासन्नतया तेन ग्रीष्मे तीव्र करा रवेः ।
देवभागे सुराणां तु हेमंते मंदतान्यथा ।।१२.४६।।

The peak of Hemant (Pre-Winter) occurs one month before 21 December, when Earth reaches apoapsis (farthest from the Sun). The peak of Grishma (Summer) coincides with periapsis (closest to the Sun).

As per Kepler's Law, the line joining the Earth to the Sun sweeps out equal areas in equal intervals of time. When the Earth reaches the location corresponding to the peak of Grishma (Summer), the angle 'r-theta' appears to cover this distance faster. Conversely, at the peak of Hemant, it seems as if the Earth's speed slows down.

It is often credited to Kepler, but long before his discovery, the same concept—**that the Earth's speed appears reduced—**was clearly stated in the ancient Surya Siddhanta. What greater proof is needed to show that our ancestors possessed profound astronomical knowledge?

If we attempt to determine the time frame when all three factors aligned, it comes out to be around 12,000 BCE. Modern astronomers argue that such events are cyclic and reoccur, which is true. However, when software is used to simulate conditions one lakh, two lakh, or even three lakh years back, no identical alignment is observed. This makes the event truly unique.

The positions of the Nakshatras and the Earth's axial tilt, as mentioned in the Surya Siddhanta, indicate a time frame of around 7500 BCE. This discovery was made by Dr. Anil Narayan, a scientist who worked with ISRO and DRDO.

The mathematics involved in calculating the Sun's location is accurately recorded in the Surya Siddhanta, which points to a period of 5300 BCE. Earlier, it was assumed there were 28 Nakshatras, but later the star Abhijit was removed from the list. The Surya Siddhanta documents this removal but does not explain when or why it occurred. The intriguing part is that the answer is provided in the Mahabharata.

The Mahabharata does not suggest that this happened during its own period but mentions that it occurred around 14,500 BCE. These references were obtained by studying other epics and various scientific disciplines.

From this, it is evident that the timelines mentioned in the Surya Siddhanta are extremely ancient—but this should not come as a surprise.

7

Determination Of The Time Of The Rigveda And The River Saraswati

Since the Rigveda is mentioned in both the Ramayana and the Mahabharata, it is evident that its period must have preceded both of them. There are no specific astronomical references in the Rigveda by which its timeframe can be approximately determined. Hence, we must rely on the era when the River Saraswati flowed, for from this we can confidently infer the time of the Rigveda.

The River Saraswati, which gradually dried up, continues to live in the consciousness of the Indian people. Even though no one—not in the present age nor for thousands of years—has actually seen the river, Saraswati still occupies a sacred place in the minds of all Indians. We shall use the evidence related to the existence of the Saraswati to determine its era and, through it, the antiquity of Indian civilization.

अम्बितमे नदीतमे देवितमे सरस्वति ।
अप्रशस्ता इव स्मसि प्रशस्तिमम्ब नस्कृधि ।।

(Rigveda 2-41-16)

The river is described as having such a powerful flow that it cuts through mountain rocks as effortlessly as a knife through lotus stems, flowing as swiftly as the fastest horses.

What Exactly Happened To The River Saraswati?

To establish any historical truth, substantial evidence is required. We must therefore first present proof of the Saraswati's existence. Geological data confirms that the River Saraswati did indeed exist and that it was active roughly 172,000 years ago.

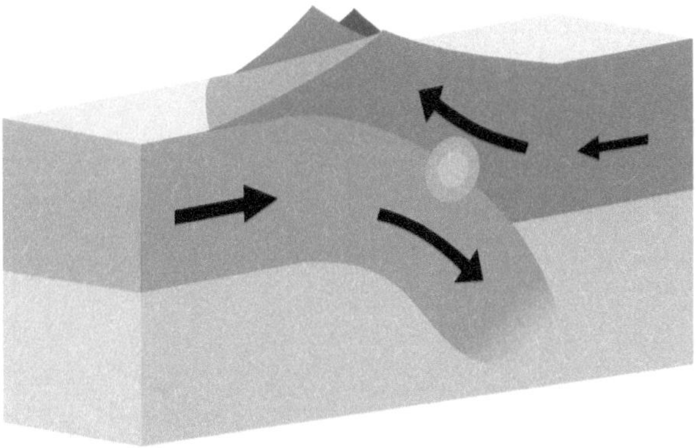

The shifiting of the Indian Subcontinent plate northward, morpho-dynamic changes took place in the rivers of India.

The northward movement of the Indian subcontinental plate caused morphodynamic changes in the river systems of India (morphology — the study of internal structure).

As shown in the diagram above, this northward shift of the Indian plate led to the redirection of several major rivers. The Yamuna and the Shatudri (Sutlej) diverged from the Saraswati, and when the Sutlej later joined the Sindhu (Indus), the Sindhu became the largest river.

The mention of the Saraswati River in the Rigveda, Ramayana, and Mahabharata has been corroborated by geological, geophysical, climatological, oceanographical, anthropological, paleontological, and remote-sensing studies.

Where did the River Saraswati originate, and where did it flow?

Around 172,000 years ago, the river flowed northward, receiving waters from the Shatudri (Sutlej) to the north, the Yamuna to the south, and many smaller rivers from Kurukshetra. It was thus a vast and celebrated river.

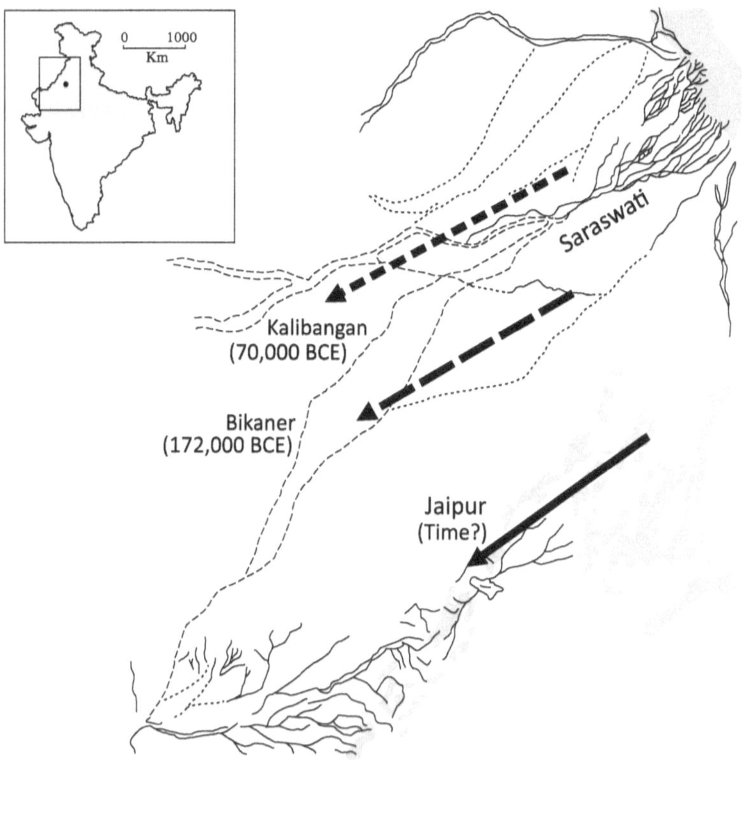

What happened later?

Around 70,000 years ago, the Saraswati—which once flowed near Bikaner—changed its course and began flowing close to Kalibangan.

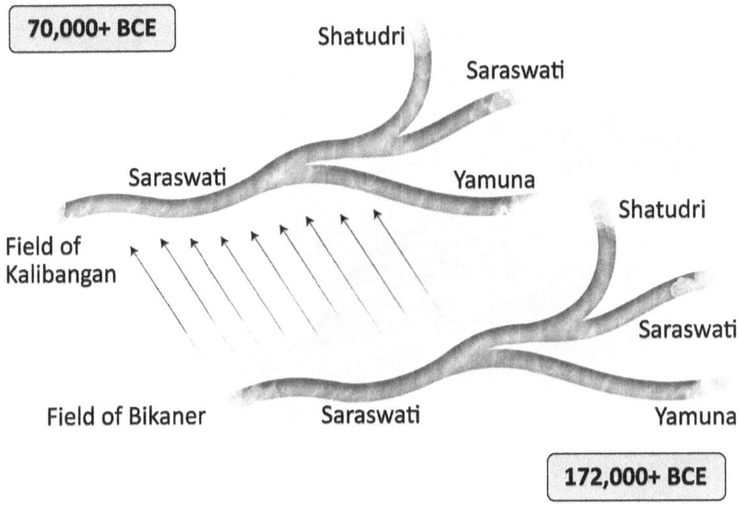

Until around 50,000 BCE, the Saraswati continued to flow with the Yamuna and the Shatudri as its tributaries near Kalibangan.

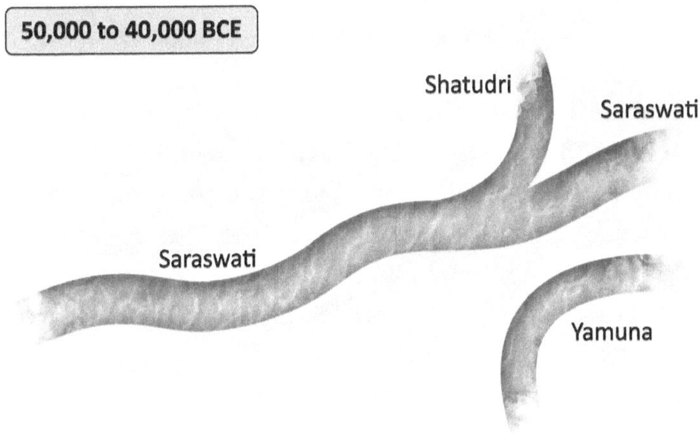

Later, between 50,000 and 40,000 BCE, the Yamuna branched off eastward (Agneya) to join the Ganga.

Subsequently, the Shatudri, instead of turning south, veered westward to join the Indus, consistent with the reference found in the Valmiki Ramayana (Ayodhyakanda, Sarga 71, Shlokas 1–2). Consequently, its waters ceased to flow into the Saraswati.

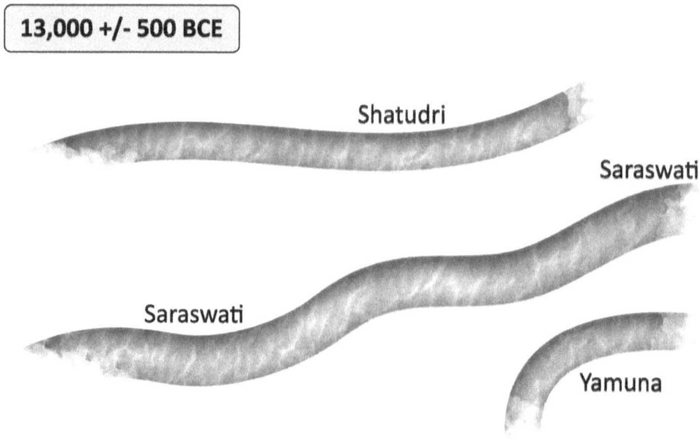

Thereafter, the Saraswati flowed only with its smaller tributaries. With the Yamuna and Shatudri no longer feeding it, the river gradually diminished in volume.

As monsoon patterns shifted, the Saraswati experienced cycles of revival and decline, eventually weakening over time.

Approximately 4,500 years ago, "after the time of the Mahabharata," the river dried up completely. Yet even after its desiccation, monsoon rains and small streams continued to flow through its ancient bed. In the form of the Bhakra, some water still runs through what was once the Saraswati's course.

But the mighty and majestic Saraswati of antiquity has vanished.

Mention Of The Saraswati River In The Rigveda, Ramayana, And Mahabharata

The hymns (Suktas) dedicated to the river in the Rigveda shed light on its grandeur. The 10th Mandala, 75th Hymn of the Rigveda, is entirely devoted to praising Saraswati.

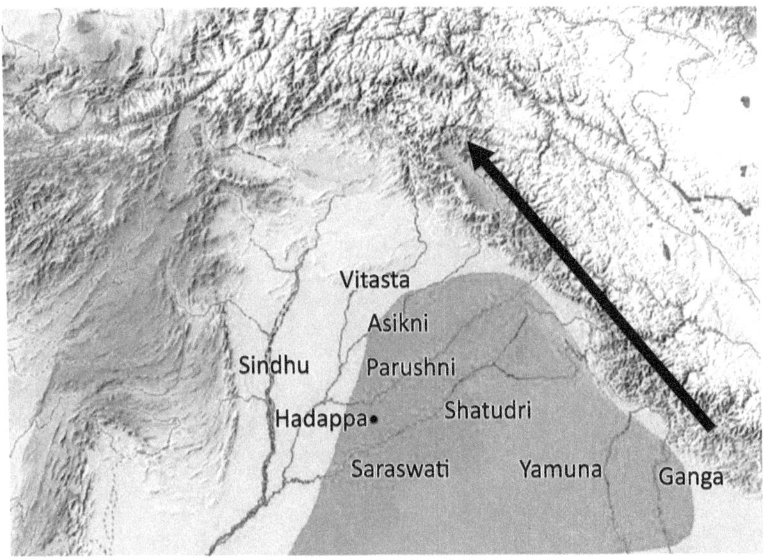

इमं मे गङ्गे यमुने सरस्वति शुतुद्रि स्तोमं सचता परुष्ण्या ।
असिक्न्या मरुद्वृधे वितस्तयाऽर्जीकीये श्रुणुह्या सुषोमया ।।५।।
तुष्टामया प्रथमं यातवे सजूः सुसर्त्वा रसया श्वेत्या त्या ।
त्वं सिन्धो कुभया गोमतीं क्रुमुं मेहत्न्वा सरथं याभिरीयसे ।।६।।

(Rigveda 10.75)

This verse mentions nineteen rivers lying toward the northwest. The order of their names is significant; if we trace them on a map from the southeast to the northwest, we find they correspond to the rivers presently in northwestern India, eastern Pakistan, parts of Afghanistan, and even extending into Iran.

The Ramayana (Ayodhya Kanda, Sarga 71, Shlokas 1–2) records that when Bharata returned from the Kekaya kingdom, he had to cross the Shatudri (Sutlej) flowing from the west. This suggests that the Ramayana cannot date earlier than 13,000 BCE.

The Mahabharata, generally estimated to have occurred between 7,000 and 4,500 BCE, belongs to a time when increased rainfall may have briefly revived the flow of the Saraswati.

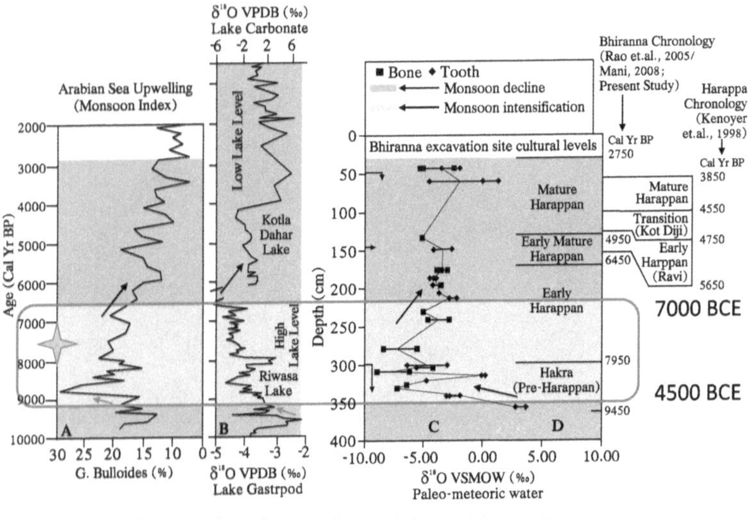

Meteorological Evidence (Anindya Sarkar 2016)

Explorers Who Searched For The River Saraswati

No branch of science is the creation of a single mind or moment. As in cosmology—where Einstein built upon the insights of Lagrange, Maxwell, and Newton—each thinker stands on the shoulders of those before him. Einstein refined Newton's laws without negating them; likewise, Kepler validated and expanded Copernicus's heliocentric model. Great discoveries arise from strong foundations, layer upon layer.

In much the same way, the search for the River Saraswati has been advanced by generations of explorers, each building upon the legacy of the last.

- Among the many explorers, Haribhau Wakankar was one who journeyed with his team from the place where the river Saraswati is believed to have flowed since the Paleolithic age. To find that source, he traversed the deserts of Gujarat and Rajasthan. The route of the river Saraswati identified by him closely matches the one determined through modern instruments. His work was truly remarkable.
- The French explorers Henri-Paul Francfort and Marie-Agnès Courty also carried out outstanding research on the river Saraswati.
- Shriniwas Kalyanramanji collected more than 8,000 seals and documented them in meticulous detail.
- Before Kalyanramanji, the explorer Parpola had also collected seals of the Sindhu Culture and recorded them systematically.
- Professor Waldia, Michel Danino, and Major General Bakshi have also conducted extensive research on the river Saraswati.

Transitions of River Saraswati		
	Scholars Say...	**Evidence Say...**
From high Himalayas to the Ocean	7000 to 4000 BCE	172,000 to 4500 BCE
Yamuna separates...	2700 BCE	50,000 BCE
Shatudri separates	2700 BCE	13,000 BCE
Intensification of Monsoon	7000 to 4500 BCE	13,000 BCE
Drying of Saraswati begins	1900 BCE	4500 BCE
Indian Antiquity	10000+ Years	200,000+ Years

The opinions of the principal academicians have been placed in the first column of the above table. Among these, apart from the fact that the Saraswati River dried up around 1900 BCE, all other statements are entirely incorrect. The time frames they propose are wrong, and that remains the central issue.

The earthquake in Kalibangan, which caused the destruction of the ancient city of Harappa about 4,700 years ago, is correctly identified. Due to this earthquake, fissures near the Yamuna increased approximately between 4,700 and 3,700 years ago. The flow of the Yamuna, which once ran westward, began turning eastward around 3,700 years ago. This implies that subtracting 2,000 years from 3,700 years gives 1,700 BCE — which is impossible. An earthquake at Dholavira about 4,200 years ago is also correctly noted. Dwarka was submerged between 3,500 and 3,000 years ago, and subtracting the current 2,000 years gives around 600 BCE. However, as we lack sufficient evidence to determine the exact date of Dwarka's submergence, 600 BCE is clearly incorrect. Historical records confirm this. The Shatudri (Sutlej), which once flowed southward into the Saraswati, began diverting westward around 2,600 years ago. Thus, subtracting 2000 years gives 600 BCE, which again is implausible. The drying of the Saraswati and the relocation of communities settled along its banks occurred around 2,500 years ago.

The Rigveda states in Sanskrit:

गिरीभ्या आ समुद्रा:

The river Saraswati, originating in the Himalayas and flowing to the sea, was active between 7000 BCE and 1900 BCE. This reasoning arises from the belief that the Rigveda was composed around the same era as the Mahabharata by Vyasa. However, that does not mean both were written simultaneously. The main composition of the Rigveda, including the Shakala Samhita, dates much earlier—over 24,000 years ago.

- The Yamuna separated from the Saraswati around 50,000 BCE (scholars, however, suggest ~2700 BCE).
- The Shatudri separated from the Saraswati around 13,000 BCE (scholars' opinion: ~2700 BCE).
- Both parties agree on a major monsoonal shift occurring between 7000 BCE and 4500 BCE.
- After that, the river Saraswati began drying around 4500 BCE (scholars' opinion: ~1900 BCE).

Both sides concur that by 1900 BCE, the Saraswati River had completely dried up.

According to mainstream academia, Indian cultural history extends back at most 10,000 years. They hold firmly to this belief, largely because their foundational assumptions are flawed, as shown earlier. Though the Ice Age occurred around 10,000 years ago, in some regions the climate was still suitable for life — and that much is correct.

Shankaracharya states in Sanskrit:
ज्ञानं न पुरुषतंत्रम्, किन्तु वस्तुतंत्रम्

He teaches that we should not believe something merely because someone asserts it, but only when supported by evidence and reason. For example, if I look out the window and see red flowers, I do not need anyone else to confirm their color. I rely on my own sight as proof. Likewise, even if the Vedas claimed that fire is cold, we would not accept it as truth. Truth, therefore, is not dependent on any one person's opinion.

Over 35,000 years of History

The Mahabharata dates back approximately 7,000 years, the Ramayana about 14,000 years, and the Rigveda around 21,000 years. To remember this, recall the table of sevens: 7, 14, 21 — and then 28. Thus, 28,000 years ago, the positions of the Nakshatras shifted.

But did Indian history stop there? Certainly not.

There exists evidence older than 35,000 years, found in the Maitrayani Aranyak of the Upanishads, where King Bruhadrath of the Ikshvaku dynasty is mentioned. It also describes a change in the pole star and the recession of the sea. The Spring Equinox occurred during the Magha Nakshatra. A similar phenomenon took place about 10,000 years ago, though the sea did not recede then. But around 35,000 years ago — 25,000 years before 10,000 years ago — all three events occurred simultaneously.

Through this, we can infer that Indian history extends back at least 35,000 years.

How Far Back Can We Trace The Existence Of The River Saraswati

A paper published by James Blinkhorn in 2020 confirms that the river Saraswati flowed near Bikaner roughly 172,000 years ago. He also noted that at Nal quarry near Bikaner in the Thar Desert of Rajasthan, fluvial activity was detected, indicating the presence of a river that flowed during 172–174, 140–150, 79–95, and around 26 thousand years ago. Evidence also suggests that after the river's flow ceased, the Ghaggar–Hakra began to flow nearby. As discussed earlier, the Sutlej (Shatudri) and Yamuna had already diverted northward away from the Saraswati around 70,000 years ago.

Fluvial activity in Thar desert near Bikaner.
Evidence of flowing water.

When the river Saraswati flowed, whether civilization existed along its banks or whether the Rigveda was composed then remains uncertain. Civilization may or may not have existed at that time. Though there is no definitive proof, as stated in the Nyaya Darshan Shastra, just as there is no evidence confirming a civilization, there is equally no proof denying it — 'Anupatya upapatibya' (Nyaya Darshan Shastra).

Yet, other forms of evidence also support its existence.

- We know that the Indian subcontinent collided with the Asian continent, and the resulting tectonic movements drastically altered river systems in northern India.
- Around the same time — approximately 75,000 years ago — the Toba volcano erupted on Sumatra Island. The eruption released massive amounts of ash, possibly affecting the Indian subcontinent and triggering significant geological transformations. While the exact link remains uncertain, the timing aligns closely.
- The northward movement of the Indian tectonic plate caused morphodynamic changes in the river systems, leading to the formation of the Himalayas.
- As shown in photo number 53, which is a map illustrating the riverbed of the Saraswati shifting from Bikaner to Kalibangan, it appears that during that time, the Saraswati flowed beneath Bikaner toward Ajmer, Jaipur, and Pushkar before turning northward. Pushkar was an important site during the Treta Yug.
- The path of the river depicted in the map was first traced by Mr. Ghosh in 1970. His research proposed this as the probable course of the river, and observations in 2020 have confirmed its accuracy.
- From about 70,000 BCE to 3,000 BCE, the Saraswati flowed through Kalibangan. Though its flow gradually diminished, it continued until around 5000 BCE, remaining substantial until roughly 2000 BCE. Thereafter, for reasons likely linked to the Last Glacial Maximum, much of its water — primarily from Himalayan sources — became locked in glaciers, reducing the river's flow. By around 5000 BCE, the Saraswati had nearly dried up.
- We have also seen that the Yamuna separated from the Saraswati between 50,000 and 40,000 BCE. Many explorers

have studied this. How, then, do we connect these evidences with our cultural heritage?

- Valmiki, in the Ramayana, mentions several times that "the Ganga and Yamuna flow together." This implies that the Ramayana could not have occurred more than 50,000 years ago.

In the Bhagavata Purana, Krishna advises not to rely on a single kind of evidence alone.

श्रुती: प्रत्यक्षमैतिहयं अनुमानं चतुष्टयम् ।
प्रमाणेश्वनवस्थानाद् विकल्पात स विरज्यते ।।

(Bhagavata Purana, Skanda 11)

Listening (Shruti), direct perception (Pratyaksha Pramana), history (Itihasa), and inference (Anumana) are the four means through which truth must be established.

In science, without explanation, prediction, and testing, we cannot claim anything to be proven. Krishna adds "Śruti" and "Itihāsa" to this framework of knowledge.

Let us now see how the explorers of the Saraswati River apply these principles.

The Signature Of The River

Every river possesses its own unique identity—its signature. From this signature, one can recognize the river unmistakably. Different regions yield distinct sands, stones, sediments, and minerals.

While an ordinary person may not notice these variations, to an expert, the differences are subtle yet unmistakable.

For example, consider the illustration below: there is River A and River B. Two different colored dots indicate their respective courses. When we excavate a dry riverbed, we find distinct layers of sediment, from which their respective ages can be determined.

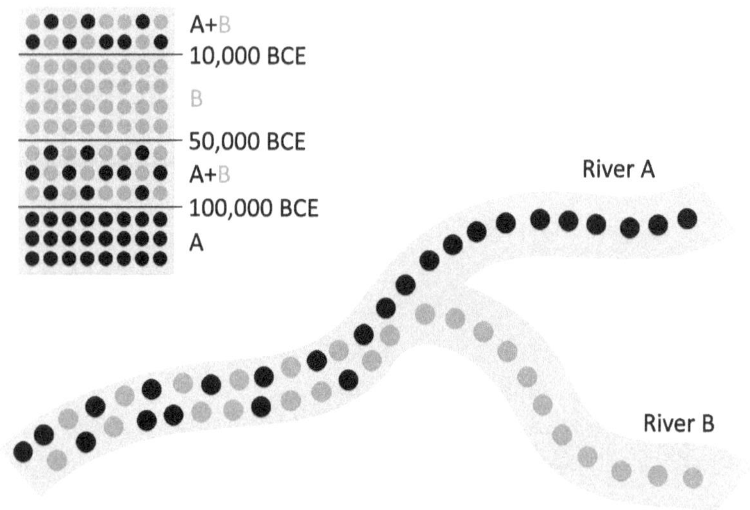

By studying these, we can understand the difference between the rivers when they flowed separately and after they merged.

Is The Source Of Perennial Rivers Always Glacial?

Many scholars claim that the source of all perennial rivers must be glacial—and that is where their error lies. Another common assumption is that between 7000 BCE and 4500 BCE, perennial rivers originated from the Himalayas and were fed by glaciers. Consequently, it is believed that during the time of the Rigveda (7000–4500 BCE), the river mentioned in the text was the Saraswati.

गिरीभ्या: समुद्रा: सरस्वती — this verse indeed belongs to that era.

If the assumption itself is flawed, the conclusion will naturally be incorrect.

A river need not draw its waters from glaciers to be perennial. In India and elsewhere, there are many great rivers that flow throughout the year without glacial origins — for example: Chambal, Betwa, Sone, Narmada, Tapi, Godavari, Krishna, Bhima, Tungabhadra, Kaveri, Vaigai, and Tamraparni.

Are only the glacier-fed rivers truly perennial? On this subject, Mr. Chatterji has written a remarkable paper, which startled several scholars.

Henry Frankfurt, Dave, Khonde, and other explorers have demonstrated that in the past 10,000 years there has been no such river that began in the Himalayas and flowed all the way to the sea. Such a mighty river did exist, but over 10,000 years ago — not within the last 10,000 years.

According to Anindya Sarkar's research (p. 60), there was heavy monsoonal activity between 7000 and 4500 years ago. This is confirmed by studies of the Arabian Sea, Rajasthan's lakes, and the Sindhu-Saraswati riverbed. During this period, the Saraswati's waters swelled due to intense monsoons.

These findings align with descriptions in the Mahabharata, which mention high water levels in the Saraswati. This indicates that the river indeed flowed during the Mahabharata period (5561 BCE)—but it would be inaccurate to equate this Saraswati with the one of the Rigvedic era.

From the Rigveda's Tenth Mandala (Sūkta 10:75)—a relatively later composition—modern scholars note that Saraswati is named among other rivers. Yet the same hymn also declares the Sindhu as the greatest.

Although the precise date of this Sūkta is uncertain, it is likely that the Saraswati was the largest river before 13,000 BCE.

Conclusion

- The Saraswati River flowed in her full splendor from 172,000 BCE to 22,000 BCE.
- Around 50,000 BCE, the Yamuna branched off from her course.
- Due to the Last Glacial Maximum, the flow of the Saraswati changed, yet she remained immense between 22,000 BCE and 13,000 BCE.
- The Shatudri (Sutlej) separated from her around 13,000 BCE.
- During the time of the Mahabharata (5561 BCE), heavy monsoons had returned.

Truth always prevails. Even if later conclusions prove our earlier beliefs mistaken, Sanātana Dharma accepts correction with humility.

According to time, place, and circumstance, our Sanātana Dharma remains ever open to new perspectives.

सनातनो नित्यनूतनः यास्काचार्य । — says Yaaska Acharya.

इति सरस्वती आख्यान ।

PART II

ANCIENT INDIAN ASTRONOMY : OBSERVATIONS, PROOFS AND CONCLUSIONS

8

Detailed Astronomical Observations By Raja (King) Bruhadrath

In both our own and Western civilizations, ancient epics contain references to great floods. However, apart from Indian epics, nowhere else is it mentioned that sea levels receded and rivers dried up. This phenomenon corresponds to the Last Glacial Maximum. Western civilization and recorded history span only about 4,000–5,000 years, and therefore cannot easily envision periods predating the Last Glacial Maximum. It is only in our ancient Indian texts that we find accounts of these events, often accompanied by precise astronomical observations.

One such example is that of Sage (Rishi) Agastya "drinking the ocean." Symbolically, this corresponds to the period of the Last Glacial Maximum. At the same time, it was observed that the star Agastya, located in the southern sky, became visible from northern India up to Kurukshetra—just as it is seen there even today.

Due to the Earth's axial tilt (obliquity) and axial precession, which cause the equinoxes to shift, certain stars that were visible only south of the Vindhya Mountains 13,000 years ago later became visible north of the Vindhyas after about the same period. This means that the southern star "Agastya" (Canopus)—though never a South Pole Star—appears to change its visibility range due to Earth's precession, and can now be seen from Kurukshetra (North India) in the sky.

Are there other examples in ancient Indian texts that mention the lowering of sea levels? Indeed, such an event is described where Parshuram commands the sea to recede, leading to the formation of Konkan and Kerala.

In the Maitrayani (Maitri) Aranyaka Upanishad, there is a passage describing a change in the Earth's axis:

अथ किमेतैर्वान्यानां शोषणं महार्णवानां शिखरिणां प्रपतनं
ध्रुवस्य प्रचलनं स्थानं वा तरूणां निमज्जनं पृथिव्या: स्थानादप्रसरणं

Through this verse, it appears that the shloka conveys the idea that the North Pole Star, which we perceive as stationary and fixed, is not truly immovable, but changes its position over time.

In the Maitrayaniya Aranyaka Upanishad, there is a dialogue between Raja (King) Bruhadrath and Sage Shakyayan. Raja Bruhadrath expresses concern over signs of receding sea levels, the movement of the Pole Star, and the collapse of mountains. Though the conversation begins dispassionately, it gradually reveals insights into time reckoning, astronomical phenomena, changes in the heavens, and decline in living lineages—that is, the deterioration of humans, animals, and plants alike.

King Bruhadrath, a ruler of the ancient Ikshvaku dynasty, is mentioned only in the Maitrayaniya Aranyaka Upanishad and the Rigveda, and nowhere else. He speaks of many kings and dynasties, yet makes no mention of Shri Ram, who is well-established in history. His narrative ends with Raja Harishchandra of the Ikshvaku line and also refers to the dynasties of Bharat, Marut, and Avikshit.

This suggests that he lived long before the era of the Ramayana—for it would have been unthinkable for him not to mention a king as eminent as Shri Rama.

His second observation concerns the lowering of sea levels, which would have exposed vast tracts of land. Another of his observations is that the Pole Star had drifted far from its earlier position.

In the Maitrayani Aranyaka Upanishad, it is stated that the year began with the Magha Nakshatra (Regulus/Leo region). The Spring Equinox occurred in Magha Nakshatra, meaning that the year once began with spring. This alignment existed around 8,000 years ago—and, before that, during the era of King Bruhadrath, as recorded.

However, 8,000–9,000 years ago, when the Spring Equinox fell under Magha Nakshatra, the sea level was rising, not falling. Yet Raja Bruhadrath laments a decreasing sea level, implying that his period predates 8,000 years ago. Such an astronomical alignment last occurred around 34,000 years ago.

From this, at least one conclusion emerges: King Bruhadrath of the Ikshvaku dynasty lived during the time of the Rigveda. In his dialogue with Sage Shakyayan, he speaks of astronomical shifts that deeply troubled him.

From astronomical reasoning, we learn about the connection between Magha (Nakshatra), Agni (Fire), the Spring Season, and the Year—their beginnings and relationships. Agni (Fire) is another name for Spring, symbolizing warmth and renewal. It is clearly noted that one sweats more during that period. This indicates that when the year began with Magha Nakshatra, this celestial configuration recurred twice like a cosmic wheel—once about 9,000 years ago, and again around 34,000 years earlier. The fall in sea level is mentioned only in connection with the latter, approximately 34,000 years ago.

Thus, King Bruhadrath of the Ikshvaku dynasty lived in a truly ancient epoch, affirming that our civilization's history reaches back unimaginably far. Even then, we possessed profound knowledge of astronomy and geography, and to record it all, we used a precise and beautiful language—Sanskrit.

9

Krutika, Ruksha And Aaraksha

It is fascinating to read about the meticulous observations and profound studies conducted by our ancient scholars on the stars, constellations, and their movements. We know by now that after some 10,000 years, our Pole Star changes. Approximately in 3,000 BCE, Thuban was our Pole Star, and the Saptarishi constellation was then located differently from where it is today. It was in the extreme north. In the present era, if we observe the Saptarishi, they rise slightly from the northeast and set toward the northwest. During the same period, Krutika rises from the east. Our ancient texts note that the distance between Saptarishi and Krutika changes over time—it either increases or decreases. Saptarishi and Krutika are mentioned together in the Shatapatha Brahmana, Ramayana, and Mahabharata.

The Saptarishi sometimes go very close to the north pole—that is, toward the north—and sometimes move away to the extreme northeast. This shift occurs approximately every 2,000 years. From these celestial observations, a story emerged, which goes like this…

In our Puranas, there is a popular legend. Mahadev's intense fire (Agni) was offered to Ganga, and from that radiant brilliance, Kumara was born. Agni placed this brightness near the six Krutikas, and Shadanan was born—he had six faces. Skanda, Kumara, Kartikeya, or Shadanan are different names for the same divine being, the son of Shiva and Parvati. The Krutikas nurtured the child and fed him milk, hence he came to be known as Kartikeya. For this reason, they are also identified as the 'Shan Matrukas' (Six Mothers). However, because of this event, the six Krutikas also became disgraced.

Our Puranas further state that the Krutikas are the wives of the Saptarishis. Believing that their wives had given birth to Skanda, six of the seven Rishis (sages) abandoned them and went away. Only Arundhati was honored as a devoted wife and remained with Vasishtha, one of the Saptarishis. This divine tableau is reflected in the night sky even today.

Saptarishi is also called 'Riksha,' meaning 'Bear,' in the Puranas. What we now call 'Dhruvamatya' or 'Dhruva Nakshatra'—the Pole Star constellation, where our North Pole Star Polaris is currently located—was also known as 'Aaraksha.' This is mentioned in the Rigveda. The 'Small Bear' and 'Big Bear' of the Western tradition

Pleiades

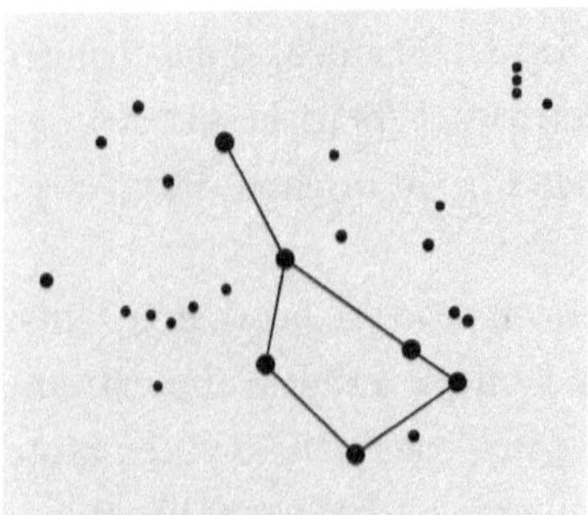

Amba, Dula, Nitatni, Meghayanti Abhrayanti, Varshayanti, Chupunika

Ursa Major

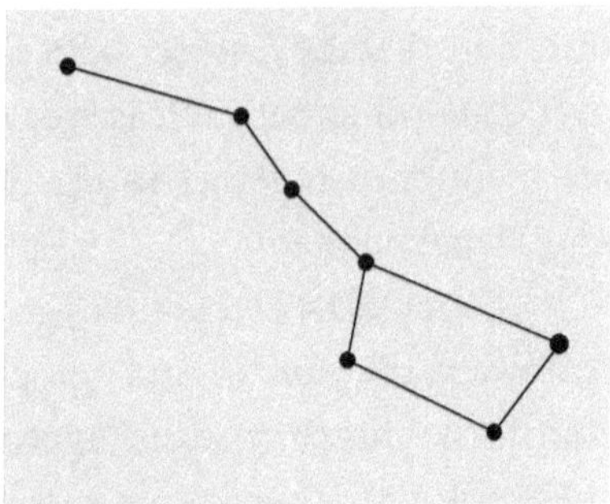

Kratu, Pulaha, Pulatsya, Atri, Angira, Vasishtha, Marichi & Arundhati

correspond to Ursa Minor and Ursa Major, meaning precisely 'Little' and 'Great Bear.'

It is also important to understand the shapes formed by the stars of Saptarishi and Krutika. If you look at the diagram, you will notice that both resemble a kite or a spoon. The only difference is that Saptarishi is larger, while Krutika is smaller. Krutika consists of six stars, while Saptarishi comprises seven stars, along with the star Arundhati.

In our Puranas, therefore, they are referred to at various places as 'Sapta (7) Matrukas' and elsewhere as 'Shan Matrukas.' In the Taittiriya Brahmana, they are described as 'Sapta Matrukas.' The Kailasa Caves in Verul also depict the Sapta Matrukas. However, in the Shatapatha Brahmana, Ramayana, and Mahabharata, six stars of Krutika are mentioned. Thus, both Sapta Matruka and Shan Matruka appear in ancient texts at different times.

Through the above discussion, it is evident that our ancestors' powers of observation and their ability to weave knowledge into stories were extraordinary—a brilliant way to pass wisdom to the common people. Truly, our ancestors were remarkable in their insight and intellect. Shatashata Pranam! (I bow to you a hundred times!)

10

Rohini Shakata Bhed – Crossing The Cart Of Rohini

Rohini Shakat Bhed is an event described in great detail and with remarkable scientific precision in our ancient texts. Let us first understand what 'Rohini Shakat Bhed' means.

Out of the 27 Nakshatra constellations, one is called Rohini, which lies in the zodiac sign (Rashi) Taurus (Vrishabha). These are essentially clusters of stars. As shown in the picture, the most prominent among them is the reddish star called Rohini, considered the 'Eye of Vrishabha (Taurus)'. The Rohini star corresponds to Aldebaran (Alpha Tauri), a bright orange-red giant star. Rohini, Elnath (Garg), and a few nearby stars of the Hyades cluster together form a pattern resembling the English letter 'V'.

This 'V'-shaped pattern marks the head of Taurus and is referred to as the 'Shakat'. Shakat means a 'Cart' pulled by an animal. Along this Shakat lies the ecliptic—the apparent path of the Sun.

The orbits of the planets Mercury, Venus, and Jupiter lie close to the ecliptic; they never stray far from this plane. Because of this, these planets rarely approach Rohini's Shakat (the 'V') and seldom cross it.

In contrast, the Moon often passes near Rohini's Shakat. The Moon's orbit is inclined by about five degrees to the ecliptic, allowing it to move close to this region. Only two planets remain that can be seen with the naked eye and wander far enough from the ecliptic—Mars and Saturn.

Ancient Indian texts tell us that when either Mars or Saturn passes through or near the Shakat, the event is called 'Rohini Shakat Bhed' (Crossing the Cart of Rohini). This phenomenon is noted in the Surya Siddhanta.

वृषे सप्तदशे भागे यस्य याम्योऽशक द्वयात ।
विक्षेपोऽभ्यधिको भिन्द्यात रोहिण्याः शकटं तु सः।।

This observation is only briefly mentioned in the Surya Siddhanta, as the epic was meant not for the general public but for scholars of astronomy. However, the Brihat Samhita by Varahamihira discusses this in greater depth, stating:

रोहिणी शकटमर्कनंदनो यदि भिनंती रुधिरो ऽ धवा शिशी ।
किं वदामी यदनिष्टसागरे जगदशेषमुपयाति संक्षयम ।।

Here, 'Rohini Shakata Arka Nandano'—where 'Arka Nandano' means the sons of the Sun, Mars or Saturn—implies that if either approaches Rohini, it is regarded as an ominous sign, unfavorable for all living beings on Earth.

Our ancestors must have arrived at this conclusion after centuries of meticulous observation. They likely noted celestial patterns that appeared before or during great catastrophes on Earth and associated such arrangements with ominous events. Or perhaps, when they observed eclipses or other rare alignments, they connected these with significant earthly happenings. Whatever the case, the brilliance and observational genius of our ancestors deserve the highest respect.

It is difficult to determine precisely the timeframe of Varahamihira, though it is estimated to be around 2,000 years ago. The Surya Siddhanta, however, is believed to have been composed roughly 40,000 years ago.

Graha Laghava says:

गणेश दैवज्ञ सोळावे शतक

भौम्याक्यौं: शकटभिंदा युगांतरे स्यात ।

सेदानीं न हि भवति दृशि स्वपाते ।।

Bhūmicha Arka means the sons of Earth—Mars and Saturn—who possess the capability of crossing the Cart of Rohini. Yet, it is stated that this event will not occur in our current age. Since it is mentioned in ancient scriptures, it must have happened in earlier times.

Now, let us explore the mention of Shakat Bhed (Crossing of the Cart).

S. B. Dixit, a great scientist recognized not only in India but globally, remains largely unknown to Indians today. Around 150 years ago, without modern software and relying solely on mathematical calculations, he determined the positions and orbits of Mars and Saturn. Through his work, he concluded that both Mars and Saturn crossed Rohini's Shakat around 5,000 years ago, approximately in 5,294 BCE.

More recently, Professor Vahiya, along with world-renowned astronomer Mohan Apte, Parag Mahajani, and Mr. Jamkhedkar from TIFR (Tata Institute of Fundamental Research), studied this event using NASA's Jet Propulsion Laboratory (JPL) data and Hubble Space Telescope archives. Their research aimed to determine how often and when this crossing might have occurred. They found that Saturn cannot geometrically align to perform such a crossing, while Mars has done so during four epochs: 9860 BCE, 9371 BCE, 9339 BCE, and 5384 BCE.

Mr. K. Chandra Hari, another Indian astronomer, redefined 'Shakat Bhed' slightly, suggesting that if a planet comes within two degrees of Aldebaran (Rohini in Taurus), it can be considered a crossing. This definition was likely introduced for convenience, but it does not qualify as true 'Rohini Shakat Bhed'. According to him, Mars can never perform Shakat Bhed, and he does not extend his analysis beyond 3000 BCE. He also disagrees with Vahiya's conclusions.

Thus, Vahiya believes Saturn never performs Shakat Bhed, while Chandra Hari believes Mars never does. Meanwhile, the rest of the world remains unaware of what Indian astronomers are even discussing!

Vahia and others say...	Amey Modak says...
9860 BCE	9862 BCE
9371 BCE	9373 BCE
5284 BCE	5286 BCE
9339 BCE	No related example found.

Ameya Modak, a modern scientist, using advanced software, demonstrated that Mars could indeed perform Shakat Bhed—passing directly through the region of the three bright stars of Taurus near Aldebaran, not outside it.

As seen in the table, there is a consistent difference of two years between the two studies. According to Vahiya, Mars crossed Rohini's Shakat in 9860 BCE, 9371 BCE, 5284 BCE, and 9339 BCE, while Ameya Modak's calculations show similar results except that no event occurred around 9338 BCE.

Vahiya suggested that in 9341 BCE, Mars underwent Shakat Bhed, but upon closer examination, Mars was outside, not inside, the triangular region. However, around 5286 BCE, Mars appeared near the star Garg, within the Rohini asterism, marking a distinct and extraordinary event. Such a rare occurrence has not repeated in the last 6,000 years.

Looking further back—around 15,000 years ago—such events may have happened at least ten times. Mars comes near the Rohini (Aldebaran in Taurus) region roughly every two years, but only rarely passes directly through the Shakat (Cart) region. Thus, in approximately 13,000–14,000 years, Mars has approached Rohini about 7,000 times but has performed Shakat Bhed only ten times.

Ameya Modak also identified a new lower limit for such events—the last occurrence around 4355 BCE. Nilesh Oak, however, found the year to be 5570 BCE.

Rohini Shakat Bhed By Saturn — Apparent Crossing Of The Rohini Region By Saturn

S. B. Dixit, deeply devoted to the Surya Siddhanta, asserted that if the text mentions Saturn performing Rohini Shakat Bhed, it must have occurred. His faith in the Surya Siddhanta was as unwavering as the 19th century's faith in Newton's laws. He believed Saturn

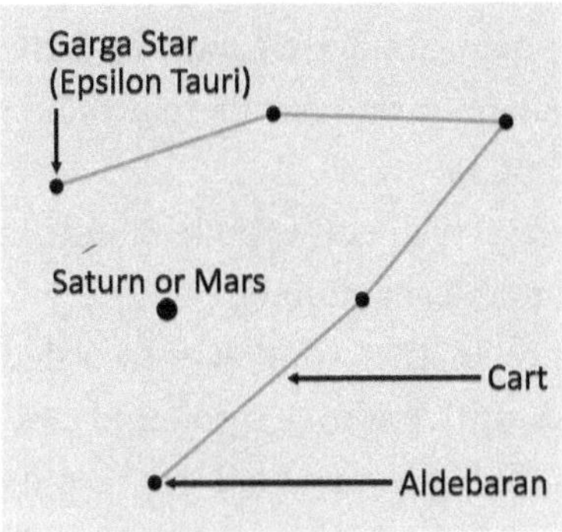

'Shakat Bhed' happened.

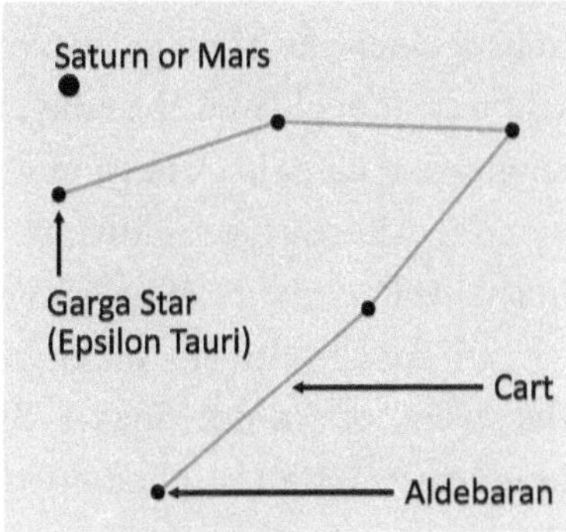

If the planet is outside the triangle of Taurus, then Shakat Bhed doesn't take place.

performed Rohini Shakat Bhed around 5000 years ago, though his tools of the time did not allow precise long-term calculations.

Chandra Hari, by altering the definition of Rohini Shakat Bhed, tried to show that Saturn had performed the crossing, but he refused to accept that Mars had done so.

Team Vahiya maintained that Saturn never performed Shakat Bhed, yet this, too, proved incorrect. Evidence suggests two occurrences—around 8105 BCE and 8046 BCE—when Saturn performed Shakat Bhed. Were these, as Varahamihira and the Surya Siddhanta imply, omens of great calamities?

Indeed, around 8000 BCE (roughly 10,000 years ago), Earth was emerging from the Younger Dryas period, a time when the Northern Hemisphere abruptly returned to glacial conditions, while the Southern Hemisphere remained comparatively warmer.

Western Indologists, however, insist that Indian astronomy is barely 1500 years old, refusing to acknowledge its ancient origins. Is it ego and intellectual arrogance that blind them? It seems difficult for them to accept that another civilization—especially India, long dismissed as backward—possesses a far older and more holistic scientific heritage than their own. Similarly, some Pro-Indic scholars, though well-meaning, hesitate to state firmly that Indian civilization dates back beyond 4000 BCE.

Yet, the Surya Siddhanta and Aryabhatiya both stand as unshakable evidence of India's astronomical antiquity.

In conclusion, no other culture in the world has recorded the event of Rohini Shakat Bhed. It remains a profound and unique astronomical phenomenon. Though many fail to grasp its significance, the fact that our ancient texts documented it is truly remarkable and worthy of deep reflection.

These researchers have examined events between 5000 BCE and 13,000 BCE, but that does not mean earlier occurrences never happened—only that no one has yet searched further. Such events likely occurred 14,000–15,000 years ago and were recorded by our ancestors. To observe, interpret, and document such celestial events and understand their effects on Earth was truly an extraordinary intellectual endeavor. These records remind us just how ancient and advanced our culture truly is—dating back perhaps 40,000–50,000 BCE.

11

The Four Dancers Of Jupiter And The Seven Hands Of Saturn

In the Mahabharata, Jupiter is mentioned and described in four different places, allowing us to trace back nearly 7,500 years. Saturn, too, is described in six different places. The following Shloka (verse) from Bhishmaparva refers to both Jupiter and Saturn.

ग्रहौ ताम्रारूणशिखौ प्रज्वलन्ताविव स्थितौ ।
सप्तर्षीणामुदाराणां समवच्छाद्य वै प्रभाम् ।।
संवत्सरस्थायिनौ च ग्रहौ प्रज्वलिताबुभौ ।
विशाखयोः समीप बृहस्पतिशनैश्चरौ ।।

(Bhishmaparva, Adhyaya 3)

To understand the meaning of this Shloka, let us start from the end.

'Bruhaspatishanaycharau' means that the planets Jupiter and Saturn are located close to Vishakha Nakshatra. The word 'prajwalitawobhau' indicates that they appear radiant, and 'Samwastarthayenau' signifies that they have remained stable for one year. 'Samwadchadayawaiprabham' means they are masking the radiance of something else—and what are they masking? That is 'Saptrarishinamudranam', meaning the Saptarishi stars. Their brilliance is so intense that the radiance of the Saptarishis seems dim before them, and there is a copper-colored plume on them ('tamrarunashikau').

(It is often difficult for many to understand the meaning of 'tamrarunashikau'. This is the beauty of Sanskrit—the uniqueness of the language lies in the fact that one word can have multiple meanings, yet only one meaning fits perfectly in its given context.)

Jupiter

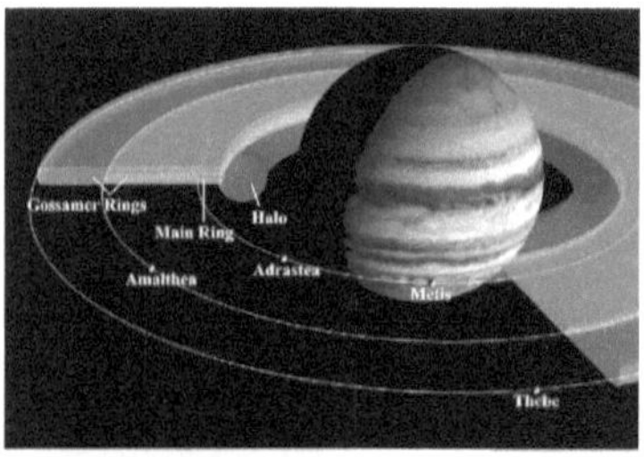

Saturn

Cap with a feather

Jupiter and its 4 Galilean moons

The word 'Shikha' can have several meanings—it can denote a spike, sharp end, projection, pointed flame, plume, peacock's crest, comb, or nipple. When we look at the accompanying image, we can draw comparisons. The last line of the Shloka refers to a period 7,583 years ago.

Let us now look more deeply into the first line. What does the word 'Tamrarunshikhou' tell us about the planets?

In the Mahabharata, there are certain astronomical observations that could not have been seen without a telescope. Saturn's rings were known in ancient times, while Jupiter's rings are extremely faint and not visible even through small telescopes. Maharishi Vyasa's use of the word 'Tamrarunshikhou' possibly refers to their rings or moons.

When we ask, "Who discovered the moons of Jupiter?" our minds immediately recall Galileo. The photograph shown here—of Jupiter with its four moons is of high quality—far beyond what Galileo could have seen. His telescope was crude. If the four moons of Jupiter appeared in a straight line, it would not be wrong to call them 'Shikha'. Today, with modern telescopes, it is easy to observe Jupiter's moons separately. But Galileo's telescope was primitive, and what he saw may have resembled a braid of light around a sphere. What Galileo claimed in the 17th century was, of course, unacceptable to the Church.

Now let us turn to the reference concerning the seven hands of Saturn. Using a telescope with a 6-inch reflector, how many moons of Saturn are visible? Usually seven of Saturn's brightest moons—a remarkable coincidence.

The truly fascinating part lies in a paper written by Colonel Pearse of the British Army, who was posted in India. During that period, many British officers, with government permission, were looting India's wealth—its food, spices, art, and ancient knowledge—and sending them to England. Despite this, Col. Pearse sent a letter to the Royal Society of London, which still exists in their archives. In that letter, he wrote that "The Indians had knowledge of four moons of Jupiter and seven moons of Saturn." He concluded that "The Indians must have possessed instruments similar to the telescope, for without such tools, they could not have made such precise observations of distant celestial bodies."

The British were well aware of the vast scientific knowledge of ancient India, yet they deliberately spread the notion that "Indians were ignorant, their ancestors were ignorant, and all real knowledge belonged to the West." This misconception took deep root, and many Indians still believe it today. The British systematically erased India's culture, traditions, and scientific legacy.

Another writer in Pearse's memoirs recounts:

"He describes a conversation between a village Brahmin astrologer and Pearse. 'I did not have time to translate everything,' he wrote, 'but the man was giving incorrect information. He said the seventh planet takes sixty years to complete one revolution around the Sun, which is completely wrong.'"

(In reality, the villager was referring to a planet beyond Saturn, since he also said it is only occasionally visible—meaning he was likely describing Uranus or Neptune, which were unknown to Westerners at that time.)

"There is no mention of Saturn's rings or moons in that account, but upon seeing the picture, I immediately recognized it as an image of Saturn. Saturn has rings and at least seven known moons. At that time, only five Saturnian moons were known, yet here six moons are depicted. It shows six hands holding each other—signifying motion within limits, unable to drift away from their planet. The seventh hand bears a crown divided into four parts, perhaps representing Saturn's rings divided into four segments."

The memoir writer further adds:

"That village Brahmin might have been referring to the four dancing girls around Jupiter."

From this, it becomes clear that the Hindus of that era possessed profound astronomical knowledge. Much of this knowledge traveled from India through Arabia to Spain, Portugal, and eventually Europe—and in that long journey, some details were inevitably lost.

The four dancing girls undoubtedly represent the four moons of Jupiter, which were unknown to Europeans at the time. However, under exceptional conditions, Jupiter's four largest moons can occasionally be glimpsed with the naked eye.

The memoir also states:

"The date on which Pearse wrote to the Royal Society was 22 September 1783. Saturn's sixth moon had not yet been discovered—Herschel discovered it on 28 August 1789. Enceladus, another moon, was also discovered by Herschel in 1789, after he completed his 48-inch Grand Telescope. Remember, all this was written by Pearse six years earlier."

Some scholars even suggest that after seeing this image once it reached Europe, Herschel might have been inspired to increase his telescope's power, enabling him to discover Saturn's seven moons. It is believed that Herschel, guided by intuition and prediction, changed his methods of observation—and succeeded.

Saturn is extremely distant from Earth, and its moons are tiny. Without a telescope, they cannot be seen with the naked eye. This clearly implies that our ancestors must have possessed some optical instrument resembling a telescope, even if not identical to those we use today.

Whenever we speak of the antiquity of Indian knowledge, skeptics demand proof. Because much of this wisdom was preserved through oral tradition, written evidence is rare. Yet the logic and coherence of this knowledge are undeniable, even if tacit. Were it not for the survival of Pearse's letter, we would have had no documentary evidence at all. For that, we owe gratitude to Col. Pearse, his anonymous memoirist, and the British Library for preserving this remarkable record.

12

Agastya, Nataraj and Apasmar

The Story of Vindhya Mountain and Agastya in Brief…

Once upon a time, the Vindhya Mountain grew resentful of Mount Meru. It felt that all the planets and stars revolved around Mount Meru while it received no significance. So, the mountain began increasing its height. It rose so high that the route from north to south became impassable. Vindhya was unwilling to listen to anyone. Then Sage Agastya, whom Vindhya considered his teacher, set out to persuade him not to increase his height. When Agastya arrived, Vindhya bent low to touch his feet, and Agastya said, "I am heading south. Until I return, remain bowed." And thus, Vindhya's height remained reduced, and the path was once again open for people to cross.

This is just a myth, but now let us look at its astronomical meaning. This can be understood when we look at the places in India from where the star Agastya was seen. Star Agastya refers to Canopus. Approximately, till around 6000 BCE, star Agastya was visible in India from Kathua village down to the southern end of India. Kathua is south of Jammu and Kashmir and north of

Kurukshetra. The star Agastya was not seen in places above Kathua. Slowly, as time passed, the star drifted farther south by degrees, until it was no longer visible from the North.

From this story about Vindhya and Agastya, we can conclude that the referenced period corresponds to around 21,000 BCE. In this, a relationship has been made between Agastya Teerth (pilgrimage site) and the Nataraj Temple. Here we will look at the expected astronomical meaning of Agastya crossing Vindhya.

For this, we are going to think by utilizing a special scientific process. We are going to find the truth of this event using the law of 'Shata Darshan' (Six Rules), which include: 'Pratyaksha' meaning Evidence, 'Anuman' meaning Inference, 'Abhaav' meaning Testimony, 'Upamaan' meaning Analogy, 'Aitihidya' meaning History, and lastly 'Artha' meaning Proof, Calculation.

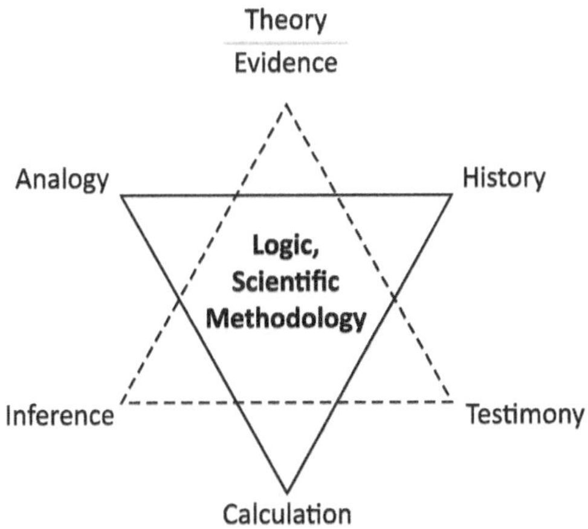

Using this method, we are going to prove that the story of 'Sage Agastya'—meaning 'Star Agastya' in the sky—crossing the Vindhya mountains reflects an astronomical event. This means that, as time passed, the star gradually shifted to lower latitudes and could no longer be seen from upper latitudes in North India.

For this method, we look for actual evidence using the Ramayana, Mahabharata, and the Puranas, which all recount the same story. From this we may infer that the Agastya–Vindhya story is an astronomical event. For written evidence, we consider an ancient astronomical text called Parashar Tantra as proof. We therefore have to look at cultural history. Why do we have Agastya pilgrimage sites at various places, and what is the relationship between Ardra Darshan Utsav and Nataraj? — these questions must be understood.

In one Shloka it is written:

न गच्छति इति अग:

Meaning: He is bound to one place and does not move.

Agam + Satyayati (collected at one place like a pile, as if a mountain.) 'Stabhnati, iti Agastyaha' is the definition of the Pole Star, which is taken as permanent. But here it is moving and continuously going forward—so what are we to make of this? The Pole Star (Southern Pole) is also not permanent. It changes roughly every 26,000-year precessional cycle. Therefore, the discussion continues in the second Shloka:

Sanskrit Shloka

अगं स्त्यायति इति अगस्त्य:

The meaning of 'Agam' here is Vindhya Mountain, and it has been stopped by Agastya at one place. Agastya is the one who has reduced the height of the mountains.

The first part had been forgotten. After that, for the second part, a very long time of approximately 12,000 years passed. Therefore, the Pole Star also changed.

There is a Shloka in 'Amarkosh':

ध्रुव औत्तानपादि: स्यात अगस्त्य: कुंभसंभव: ।
मैत्रा वरूणिरस्यैव लोपामुद्रा सधर्मिणी ।।

Meaning: Linguists have understood that 'Dhruv' is the son of Raja Uttanpad. They say that the star being talked about is the Pole Star, while the others are ordinary stars. Linguists have not been able to find the meaning of 'Lopamudra', 'Kumbhasambhava', 'Mitra, Varun', etc., nor their interrelationship here.

To know that, we will see ahead how this occurs.

We all presently know about Earth's precession speed. The speed of the Earth's axis for the entire circle to be completed is 26,000 years. Now what is the relationship of this precession speed with the star Agastya? In the diagram, the location and time of the Agastya star whenever it is visible have been indicated. Agastya is always seen from Chidambaram. Canopus has, in fact, been visible from most of peninsular India throughout the entire precessional cycle. In 2000 CE, Agastya can be seen till Kathua, and 24,000 BCE it was seen till Kurukshetra. It is thought that in between this time Agastya was not seen from anywhere in all of India, which is not exactly correct. We also have to assume how high above the horizon Agastya should be to be seen. It was taken that Agastya should be 8 degrees to 5 degrees above the horizon, then only it will be spotted. But as said by Hipparchus, he observed the star Agastya (Star Canopus) very close to the horizon, that is, at a height of 1 degree 16 minutes. Hence, it would not be correct to say Agastya was not seen.

There is also a mention of Chidambaram here, and we will see what its connection is. When uttering 'Chidambaram,' the first thing that comes to mind is 'Nataraj'. Nataraj is the God of Dance. In this temple, Nataraj is shown dancing with the demon Apasmar at his feet. 'Apasmar' means Asur (demon). How is this related here?

For that, let us first know about Mruga (Orion) Nakshatra. If we superimpose the idol of Nataraj over the picture of Mruga Nakshatra, we will understand many things. The idol of the Apasmar demon at the feet of Nataraj is also similar to the constellation Lepus, which lies just below Orion. Lepus means 'Hare'. In the sky, we can easily identify the Mrugashirsha Nakshatra (Orion). The three stars around the midsection of Mrugashirsha Nakshatra immediately catch the eye. From these three stars, we can easily identify other constellations around Mruga. Once we superimpose the image of Nataraj on the Mrugashirsha Nakshatra, we see that when we draw a straight line from his left foot, there is a bright star—Sirius. We call this star 'Vyadha' or Canis Major or Indra. In some stories, it is said that Vyadha hunts Mrug or is hunting Mrug. But in this account, the opposite is described—Mrugashirsha is hunting this star.

According to Greek mythology, Orion is the hunter and Vyadha is the dog. Once Mrugashirsha is known, we can see various other stars and constellations around it, like the two stars Castor and Pollux, whom we know as 'Diti' and 'Aditi' in Mithun Raashi, Punarvasu, Saarthi constellation, Rohini, Krutika, etc. Similarly, if we follow Nataraja's Gaja Hasta Mudra (hand gesture symbolizing an elephant's trunk) straight toward the south, we see a bright star. This star is Agastya or Canopus. The moment we superimpose Nataraja's figure on Mrug Nakshatra, the outline of Apasmar immediately superimposes on Lepus. In the temple of Nataraj in Chidambaram, Lepus, which is at the feet of Nataraj, is called 'Muyalak' in Tamil. In Tamil, 'Muyalak' again means 'Hare'. Apasmar's idol is shown with a hand mudra like 'Anjuli' or 'Anjali' or 'Onjal' (cupped hands). Does this remind you of something? It evokes the story in which Sage Agastya drank the entire sea with one cupped hand. Malabar sailors used this constellation of Rudra as the hunter and Apsmar as the demon for sea navigation. Malabar sailors call this constellation "Kuthu" or "Kutton". This information reached the Greeks unchanged, and hence in their star charts they also named the constellation Orion the hunter with a hare at his feet—though no one can explain why the hare is there.

In Mrugshirsha Nakshatra, the main verb is Mrugayati, which means "following Mruga to hunt it." Whom is he hunting? It is Canis Major. This means that this is 'shwan', Indra's dog, or perhaps Indra himself. It is the brightest star. Thus, Mrug, Lepus, and Agastya are all interconnected.

Now let us see how and when Agastya traveled to the south. While traveling south, it slowly disappeared from the horizon at successive southern locations. It was seen from above Kurukshetra—that is, Kathua—till 26000 BCE. It was not seen in latitudes above Kathua. Agastya, while traveling south, was crossing the Vindhya ranges around 19000 BCE. Here 'Agastya' refers to the star. In contrast to this, K. D. Abhyankar has given his opinion that Sage Agastya, while going south near the Vindhya mountains, noticed the Agastya star for the first time. Hence the star was named 'Agastya'. But in reality, there is no mention anywhere of Sage Agastya ever having traveled from the north to the south. We find the story, but no firm evidence or reference. Agastya star used to rise and set in the south. It has been mentioned in Parashar Tantra, an epic on astronomy, that only a star can rise and set—not a sage.

Agastya star was not seen from the latitude of Chidambaram from 13000 BCE for about 4,000 years. It was not seen above a latitude of 10.5 degrees but was visible at latitudes south of Chidambaram. As described in the Ramayana, during that time (11000 BCE to 13000 BCE), Agastya was seen from Mount Mahendra.

The three semicircular diagrams below show the location of Mruga Nakshatra as seen from Kurukshetra. Slowly, we see the Mrug Nakshatra, and also Agastya, moving gradually toward lower latitudes.

(Apasmar is a dwarf demon considered immortal in the Puranas. Apasmar is also a term for a kind of brain disorder—'seizure'. Another meaning of 'Apasmar' is 'ignorance' or 'forgetfulness'. Knowledge in the world should not be lost; therefore, Apasmar cannot be killed, but he can be kept under control. Hence Shiva, as Nataraj during the Tandav Nritya, captured Apasmar and kept him beneath his left foot. To kill him would mean gaining knowledge without any difficulty, and therefore he is not slain.)

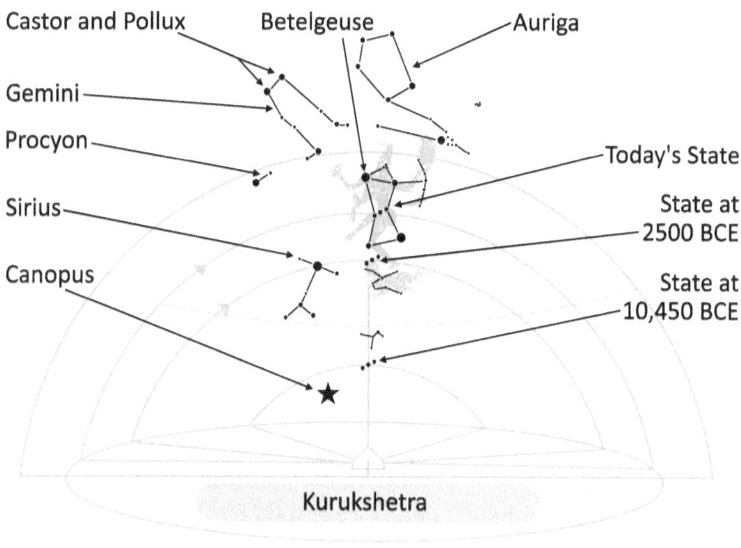

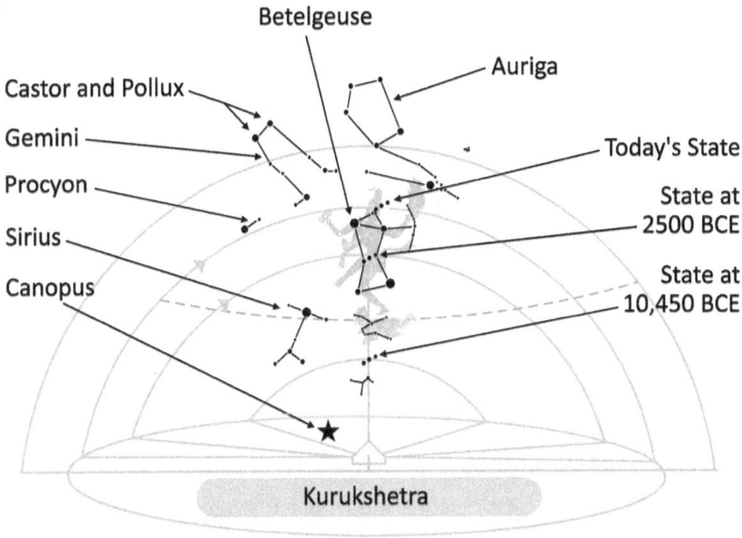
Betelgeuse
Auriga
Castor and Pollux
Gemini
Today's State
Procyon
State at
2500 BCE
Sirius
State at
10,450 BCE
Canopus
Kurukshetra

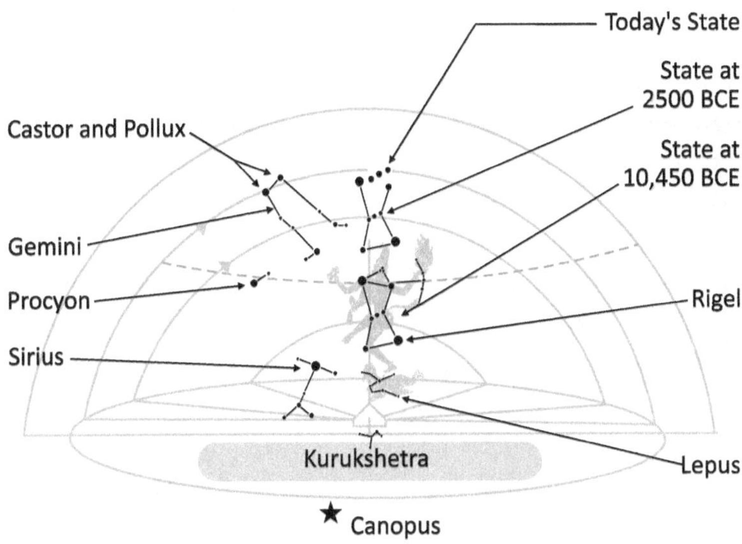
Today's State
State at
2500 BCE
Castor and Pollux
State at
10,450 BCE
Gemini
Procyon
Rigel
Sirius
Kurukshetra
Lepus
Canopus

In Tamil Nadu, the Ardra Darshanam festival is celebrated. When Ardra in Mruga Nakshatra rises—Ardra meaning the shoulder of Shiva (Rudrasya Bahu)—and Poornima come together, it is believed that on that night, Shiva in the form of Nataraja performs the Tandav. During this time, he captures Apasmar (ignorance) and keeps him under control. The Malabar sailors used this iconography for direction, like a compass, in sea navigation. They also used it for timing, for when Ardra rises, after some time Vyadh (identified with Sirius) will rise, and subsequently one can determine when Agastya (the star Canopus) will rise. Using this method, the correct southern direction becomes known, and the sea voyage can begin. Since ancient times, our people have undertaken sea voyages. Malmani Pothi from Gujarat even describes different forms of waves, resulting situations, wind speeds, and many such details. Using this knowledge, Vasco da Gama was able to locate the Indian coast. Many places were discovered by the Kutch and Malabari boatmen based on the knowledge recorded in this book. Chidambaram and Udagmandalam are both on the same latitude, but one lies in the east and the other in the west. Agastya (Canopus) becomes visible from this latitude only when it moves upward from here around 9,500 BCE. It took until 6,000 BCE to reach this point. Sri Krishna says this has formed a good Agastya teerth, a place now known as Somnath. The Agastya star is seen exactly at the height just above the Shivling in Somnath, but since a temple has now been built over the Shivling, the Agastya star is no longer visible. This is how Agastya began its northward journey, and in our time it has not yet reached as far north as Kurukshetra or Kathua. Due to Earth's precession, once again it will later form the southern pole star, and the cycle will continue. (Kurukshetra is the last stop; above this latitude it is not seen, hence its importance. Ujjain, lying on the

Tropic of Cancer, is also of great significance. From here, the sun's southward journey begins.)

रुद्राणमपि यो रुद्रः प्रभा प्रभावतामपि ॥२८॥

Ardra Nakshatra = deity Rudra

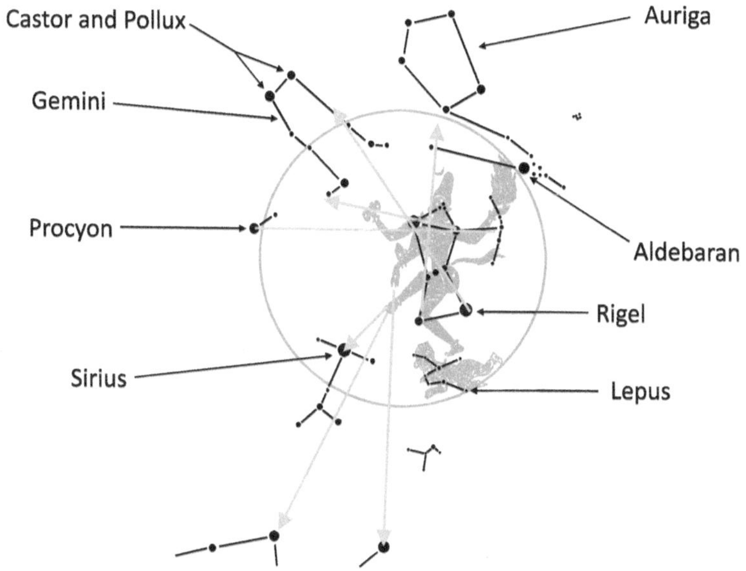

From all this above discussion the conclusion we reach is that the Vindhya–Agastya legend is actually an astronomical event. This event occurred during 21,000 BCE, demonstrating that since ancient times Indian astronomical knowledge was well advanced. The meticulous observations made by our ancestors—recorded even in puzzle-like forms—truly deserve a hundred salutations.

13

Shishumar Chakra (Wheel)

Shishumar Chakra has been described in the epic Shrimad Bhagwat, First Kand, Twenty-third Skanda. In this description, the distance between the Saptarishi, Pole Star, Sun, Earth, Moon, and other planets has been mentioned and given in Yojana. Yojana is a unit of distance whose value varies across different periods, so without going into the mathematics of these units, we will focus on the remaining discussion.

At some places, the constellation Kaleya has been mentioned as Shishumar, while at other places they have been treated as two different constellations. About 5,000 years ago, Thuban was our pole star, but it lies in the constellation Kaleya. Kaleya runs through Saptarishi and Dhruvmatsya, which are Ursa Major and Ursa Minor respectively. Therefore, Kaleya and Shishumar are distinct.

Shrimad Bhagwat says that 'Avyaktgati bhagwan kal', i.e., the one whose speed cannot be determined, yet whose presence is certain. With time, planets and constellations move, and to support them a celestial abode was assigned. Notice the precision of the

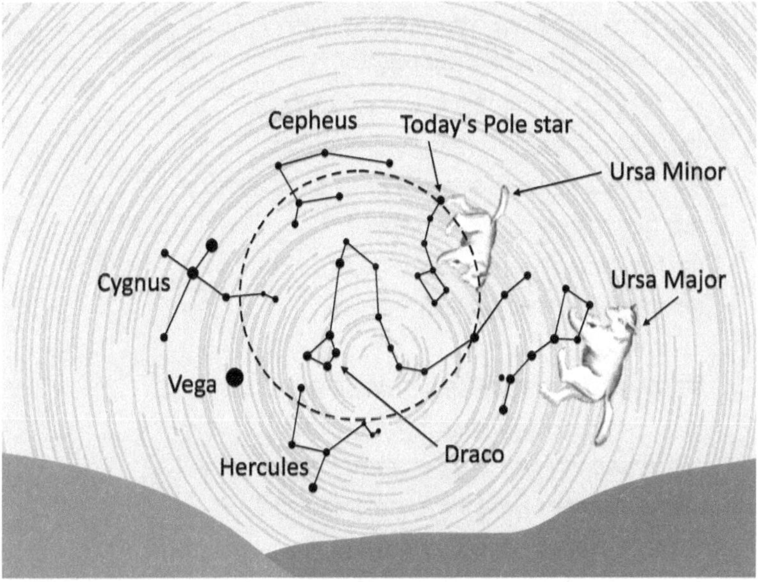

expression used. Modern science shows that constellations appear to revolve around the pole. To support the Nakshatras, poles have been designated. Just as officers change, so too does the Pole shift over time.

Further it is said that just as small, medium, and long ropes are tied to a bull to make it move within a fixed circle while grinding grain, similarly all planets and constellations move in their respective inner and outer orbits around the central pole. This analogy presents a remarkably accurate picture.

केचनैतजज्योतिरनीकं शिशुमारसंस्थानेन ।
भगवतो वासुदेवस्य योगधारणायामनुवर्णय ।।४।।

Meaning: In Bhagwat this astronomical wheel has been named Shishumar, also described as an imaginary aquatic animal. Its body has been described as circular. Shishumar is depicted coiled from the right side.

यस्य पुच्छाग्रेऽवाक्शिरसः कुण्डलीभूतदेहस्य ध्रुव
उपकल्पितस्तस्य लाङ्गूले प्रजापतिरग्निरिन्द्रो
धर्म इति पुच्छमूले धाता विधाता च कट्यां सप्तर्षयः ।
तस्य दक्षिणावर्तकुण्डलीभूतशरीरस्य यान्युदगयनानि
दक्षिणपार्श्वे तु नक्षत्राण्युपकल्पयन्ति दक्षिणायनानि तु सव्ये ।
यथा शिशुमारस्य कुण्डलाभोगसन्निवेशस्य
पार्श्वयोरुभयोरप्यवयवाः समसंख्या भवन्ति ।
पृष्ठे त्वजवीथी आकाशगङ्गा चोदरतः ॥ ५ ॥

पुनर्वसुपुष्यौ दक्षिणवामयोः श्रोण्योरार्द्राश्लेषे च
दक्षिणवामयोः पश्चिमयोः पादयोरभिजिदुत्तराषाढे
दक्षिणवामयोर्नासिकयोर्यथासङ्ख्यं श्रवणपूर्वाषाढे
दक्षिणवामयोर्लोचनयोर्धनिष्ठा मूलं च दक्षिणवामयोः
कर्णयोर्मघादीन्यष्ट नक्षत्राणि दक्षिणायनानि
वामपार्श्ववङ्क्रिषु युञ्जीत तथैव मृगशीर्षादीन्युदगयनानि
दक्षिणपार्श्ववङ्क्रिषु प्रातिलोम्येन प्रयुञ्जीत शतभिषाज्येष्ठे
स्कन्धयोर्दक्षिणवामयोर्न्यसेत् ॥ ६ ॥

उत्तराहनावगस्तिरधराहनौ यमो मुखेशु चांगारकः
शनिश्चर उपस्थे बृहस्पतिः ककुदि वक्षस्यादित्यो
हृदये नारायणो मनसि चंद्रो नाभ्यामूशना
स्तनयोरश्विनौ बुधः प्राणापानयो राहुर्गले केतवः
सर्वांणगेशु रोमसु सर्वे तारागणाः ॥७॥

Meaning:

To the right of Shishumar there are fourteen Nakshatras starting from Abhijit to Punarvasu Nakshatra of Uttarayan (Northern movement of sun towards summer solstice). To its left are fourteen Nakshatras starting from Pushya Nakshatra to Uttarashadha Nakshatra. At the back, on the right side lies Punarvasu Nakshatra and on the left lies Pushya Nakshatra. Behind it and at its feet, on the right is Ardra and on the left Aashlesha Nakshatra.

In its right and left nostrils are Abhijit and Uttarshadha Nakshatras; its right and left eyes are Shravan and Purvashadha; and its right and left ears are Dhanishta and Mool Nakshatras. Eight Nakshatras of Dakshinayan (the Sun's southern movement toward the winter solstice), beginning with Magha, occupy its left ribs, and in reverse order—starting from Mrugashirsha—lie the eight Nakshatras of Uttarayan in its right ribs. On its left and right shoulders are the Shatatarka and Jyeshtha Nakshatras.

On its chin, the upper part corresponds to Agastya and the lower to Yama; inside its mouth is Mars; the genital region is Saturn; the hump is Bruhaspati; the chest is the Sun; the heart is Narayana; the head is the Moon; the navel is Shukra; the breasts are Ashwinikumar; between Pran and Apan is Mercury; Rahu is at the neck; the entire body is Ketu; and in every part of its body lie all the stars.

Does this detailed description not align remarkably well with the ecliptic?

In the form of an imaginary aquatic animal curled upon itself, Shishumar, the entire Nakshatra Chakra (Wheel) is created around the ecliptic.

The ecliptic is the path the sun takes across the sky; this is what constitutes the Nakshatra Chakra.

14

Jyotirlinga In Central India And Sampat (Equinox)

Jyotirlinga is a composition of two words—Jyotir (Astronomical Fire, Stars, Constellations) and Linga—which are combined into a single term.

From the Puranas there is a story.

For the marriage of Shiva-Parvati, everyone had gathered near the Himalayas, and as all the people assembled there, the Earth began to lose balance. Mahadev (Shiva) then told Sage Agastya, "Only you can handle this imbalance. Go to the south so that the weight on the Himalayas decreases. Only you can do this. Please do something."

From this story we learn one thing: there was a path by which Sage Agastya began his journey southward. Along the path he took, there are many Teerthas (pilgrimage sites). (Teerth also means to go to a place by walking.) Out of the twelve Jyotirlingas in India (it is said there are 64 in all), most lie in central India along a single track, as if positioned to maintain balance. But how is this balance maintained? There must have been important events at these locations, and hence the Jyotirlingas exist there.

It is important to note that along this path lie significant places like Ujjain, Avantipuri—also called Avanti. Here stand Mahakaleshwar, Onkareshwar, and other important Shiva temples. The prime meridian, which now passes through Greenwich, is believed to have earlier passed through India, specifically through Ujjain.

Let us observe, across this important middle belt, how sunrise and sunset occur, their directions, and their relationship with the Shiva linga.

It is only during the Spring and Autumn Equinoxes—the only two days—when the sun rises exactly in the east and sets exactly in the west. On other days, sunrise and sunset are not perfectly aligned to the east and west. The sun's path across the sky is also oblique. This is due to the Earth's tilted axis, which is at an angle of 23.5 degrees. The Earth's ecliptic plane makes an angle of 23.5 degrees to the celestial equator. From spring equinox to autumn equinox, the sunrise and sunset occur in the northeastern and northwestern parts of the horizon; thus, there is summer in the northern hemisphere and winter in the southern hemisphere. In the northern hemisphere, we experience Dakshinayan Arambh (the start of the sun's southern movement). After reaching its maximum northern point, the sun turns southward—this is 'Dakshinayan Arambh.' After the autumn equinox, the sun rises and sets in the southeastern and southwestern regions of the horizon.

Astrology (Astronomy) depends on the sun. The sun reaches its maximum northern point on 21 June (Summer Solstice) and then begins its southward journey. From there, the sun slowly descends.

During the spring equinox, the sun rises from the east at 0 degrees latitude (the Equator). Slowly, the point of sunrise moves northward, and on Dakshinayan Arambh (summer solstice), the sunrise is at its maximum northern point at 23.44 degrees latitude, near the latitude of Ujjain in India. After this, the sun begins its journey southward and, upon reaching 0 degrees latitude, the autumn equinox occurs. From there it moves further south, reaching its lowest point in the southern hemisphere on the start of Uttarayan Arambh (Winter Solstice), when sunrise occurs at 23.44 degrees south latitude.

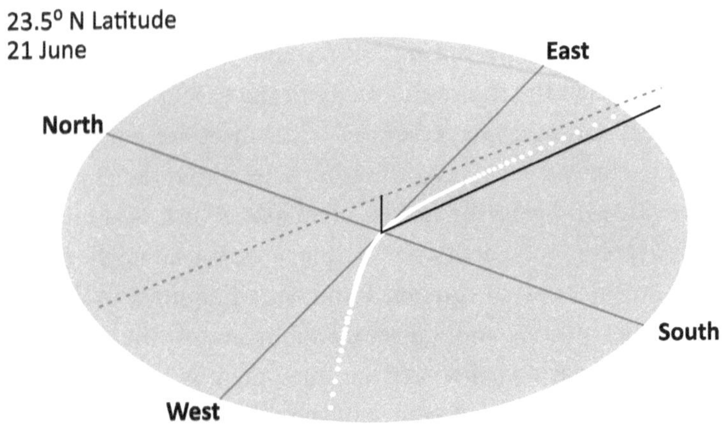

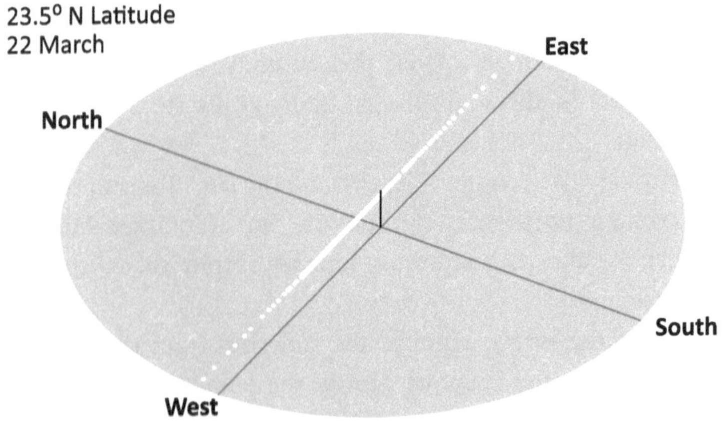

Let us observe how, during the spring equinox, the sun rises and how its shadow falls on Earth. Sunrise is from the east. As shown in the diagram, imagine the circle as Earth. At its center on the diameter, a cone has been erected. The animated progression of the shadow is shown. Only on the days of the equinox does the shadow fall in a straight line. This shadow falls slightly away from the cone, and this distance corresponds to the latitude of that location. On other days, the shadow shifts from the center and does not fall in a straight line. In the diagram, the latitude is 23.5 degrees north, as we are considering Ujjain. If we erect the same cone in Delhi, the shadow will fall at Delhi's latitude, approximately 28 degrees. When the shadow falls straight, it means the sunrise is exactly in the east and sunset exactly in the west. Thus, the precise north and south directions can be determined. This day is important for any work related to Vastu (architecture). We get only two such days. Hence, since ancient times, architectural work would begin during the spring or autumn equinox.

We have read that during equinoxes, the sun's shadow falls on Earth in a straight line. This principle is used in the Padmanabham temple in southern India. During sunset, the sun descends in a straight line. In the Gopuram (ornate entrance tower) of Padmanabham, the sun moves down at sunset in a straight line visible through the windows built into the Gopuram. Many people visit the temple during the equinox to witness this.

In many temples of Mahadev, windows are built on the roof of the Gopuram through which sun rays fall on the Shiva Pindi (a physical cylindrical symbol of the divine in the universe associated with Lord Shiva). In times when temples did not exist, people would determine the time, month, and season based on the shadow cast by the Pindi.

Now let us observe how the sun's shadow falls and where it lies when it reaches its extreme at the Tropic of Cancer (approximately

Padmanabh Swami Temple on the day of Equinox

23.5 degrees latitude) in the north (summer solstice). The cone in the center we take as the Shiva linga. As shown in the diagram, the shadow falls away from the central line.

In the Mahakal temple, during the sun's position at the extreme north, the shadow of the Pindi can become very short or nearly disappear around local noon if the temple lies close to the Tropic of Cancer. This phenomenon is truly astonishing. If there were no roof on the Mahakal temple, then during the time of the sun passing directly over the Pindi (which occurs only at locations exactly on the Tropic of Cancer), there would have been no shadow at all. Thus, when the first Pindi was erected, the disappearance of the shadow would have amazed the people, and the place would naturally have gained increasing importance.

Correspondingly, erecting such monolithic structures is an age-old village tradition. At the edge of the forest, the stones erected by our ancestors were regarded as gods, and seeking their permission before entering the forest is a custom still followed by many tribal communities. Perhaps this is how the tradition of stone-erection first began. Stonehenge in England has been proven to have both religious and astronomical significance. Similarly, in our culture, the tradition of the Jyotirlinga may also have arisen for astrological reasons.

Now let us understand the Surya Mandirs (Sun Temples) erected at or above the Tropic of Cancer. On the 23.5 degrees latitude, or the Tropic of Cancer, there is a temple at Modhera in Gujarat. The temple in Modhera is exactly on the northern latitude of 23.5835 degrees. Since Modhera lies slightly north of the Tropic of Cancer, the sun does not pass exactly overhead, but around the summer solstice the noontime shadow becomes extremely small. In this way, the temple itself functions as a natural sundial. Everyone should surely see it on the day of Dakshinayan Arambh.

The Sun Temple in Konark is on the Northern latitude 19.8876 degrees. Due to this, the sun's shadow falls on the chariot wheel; as the sun moves, the shadow moves, and from this phenomenon we can determine the time. It also allows us to observe whether the days are lengthening or shortening.

Sun temples in the Tropic during Equinox

The earth's axis, which is now tilted at an angle of 23.5 degrees, was tilted at 22.7–24 degrees in ancient times; Earth's axial tilt naturally oscillates between about 22.7–24 degrees over a 40,000-year cycle. These changes occurred approximately 40,000 years ago. In this context, several important observations and questions arise.

Was the temple at Mandhata built when the axis was tilted by about 22.5 degrees?

Were they, by erecting Shiva ling at different places, trying to mark the axis?

Were they placing the Shiva ling at locations where the noontime shadow became very short or nearly vanished?

Mahakal temple is at about 23.5 degrees; therefore, was the Mandhata temple in Madhya Pradesh built during a time when the earth's axis was tilted by 22.5 degrees?

Did our ancestors move the Shiva ling from one location to another to mark the shifting axis of the Earth?

Were they carefully observing such subtle variations?

To elucidate, we can consider an example from the Middle Ages.

Raja Sawai Jai Singh, around 1700 AD, established an observatory at Ujjain in Madhya Pradesh. The sundial there no longer shows the correct time, but when it was built, it would have been accurate. This is because, over 300 years, extremely small deviations have accumulated, rendering the current readings inaccurate. Adjustments must therefore be made according to present conditions. Hence, it is important to relocate the observatory to a more suitable place. Raja Sawai Jai Singh himself had relocated ancient time-keeping instruments, as even during his period they were no longer showing the correct time. They had to be reconstructed.

During the equinox, the shadow forms a straight, stable line, but on the solstice, due to the earth's tilted axis—even though the axis itself does not change—the sun's shadow keeps shifting. Thus, the place where the shortest noontime shadow occurs also changes over time.

King Mandhata belonged to the very ancient Ikshvaku dynasty. He lived long before Shri Ram. As per new discoveries, the Ramayana occurred 14,000 years ago, and King Mandhata lived much earlier than this. Therefore, it is natural that the temple built in his name would also be extremely ancient, constructed at a time when the earth's axis lay somewhere within its long-term natural range rather than fixed at any specific value such as 22.5 degrees.

Thus, the temple would have been erected at that location accordingly.

Part III
THE KNOWLEDGE TRADTION DURING THE MEDIEVIAL AGES

15

Vishwarupadarshan Or Was It 'The Universe' In Dnyaneshwari?

Since the last 500 years, there has been a rapid advancement in modern science. With the help of science and technology, human life has become considerably easier. In the West, it was Copernicus who ushered in modern science. Hence, it is also called the "Copernican Revolution."

What exactly is the Copernican Revolution? The 15 years preceding Copernicus are known as the "Dark Period" in Europe, as during that time there was no advancement in the field of science.

The revolutionary idea put forward by Copernicus was that the Sun is stable and all the planets in the solar system move around it. Much significance has been attached to this discovery. In this regard, I would specifically like to cite a parable written by Sant (Saint) Dnyaneshwar in the Dnyaneshwari.

आणि उदो-अस्ताचेनि प्रमाणे ।
जैसे सूर्याचे न चलता चालणे।
तैसे नैष्कर्म्यत्व जाणे । कर्मेंचि असता ।।

Saint Dnyaneshwar says — “The Sun rises in the morning, comes overhead in the afternoon, and sets in the evening, yet in reality it remains unmoving.”

One thing has to be understood: Copernicus was from the 16th century, while Dnyaneshwar was from the 13th century!

One more important aspect is the description that Saint Dnyaneshwar has used to describe the Sun. This means that in India even the common man was aware in the thirteenth century that the Sun was stationary. Now let us look at one more illustration.

जेथ उन्मीलन होत आहे दिठी ।
तेथ पसरती आदित्यांचिया सृष्टी ॥१४१॥

(Dnyaneshwari Adhyay, 11)

“Wherever the supreme vision of Vishnu falls, there the radiance of the Sun spreads.”

If we look at this line mentioned above from the cosmological viewpoint, there are over three thousand confirmed planetary systems known in our galaxy. These must be called “multiple solar systems.” In this infinite universe, countless such planetary systems must exist.

Can it be that this is what Arjuna saw?

इया मूर्तीचिया किरीटी । रोगमूळी देखे पां सृष्टी ।
सुरतरु तळवटी । तृणांकुर जैसे ॥१४८॥
वाताचेनी प्रकाशे । उडतां परमाणू दिसती जैसे ।
भ्रमत ब्रह्मकटाह तैसे । अवयव संधी ॥१४९॥
एथ एकैकाचिया प्रदेशी । विश्व देख विस्तारेशीं ।
विश्वाही परौते मानसीं । जरी देखावे वर्ते ॥१५०॥

"Arjuna, just as thousands of grass shoots grow at the base of the Kalpataru (wish fulfilling tree), so too, in my universal form, behold the multitude of universes rising like the roots of my hair."

"Just as tiny specks appear to swirl in the wind when illuminated by light, similarly, in my universal form, numerous universes move along my joints."

"Behold the world in detail, in every region of my cosmic form—and if you wish to perceive what lies beyond it, I offer no resistance."

'Laniakea' refers to a supercluster containing many galaxy groups and clusters that are gravitationally influenced along large-scale cosmic filaments. The galaxies within it are also not stable but move within the expanding universe, much like what is described in the 149th line.

हे अवतार जे सकळ । ते जिये समुद्रींचे कां कल्लोळ ।
विश्व हे मृगजळ । जया रश्मीस्तव दिसे ॥१८०॥

All these manifestations are like waves on an eternal ocean, and under the glow of that eternal light, the world appears like a mirage.

म्हणे केवढें गगन एथ होतें । तें कवणें नेले पां केउतें ।
ती चराचरें महाभूतें । काय जाहली ॥१८८॥

Arjuna asked, 'Such a huge sky was here—who took it away, and where? Where did the mobile and immobile things and the great elements go?'

When an astronaut leaves the earth and goes into space, where the earth's atmosphere ends, he naturally feels as though the sky has disappeared. So for Arjuna, in order to see the universal form, he would have had to rise beyond the earth's atmosphere and look back, and hence he would have felt like saying this.

दिशांचे ठावही हारपले । अधोर्ध्व काय नेणो जाहले ।
चेइलिया स्वप्न तैसे गेले । लोकाकार ।।१८९।।

'There was no sense of direction. Who knows where up and down have gone? Just as a dream vanishes on waking, so too does the very shape of the world dissolve.'

We do not experience a loss of direction on earth, but in space this becomes entirely reasonable. This is an important scientific truth.

'As the sun's brightness causes the moon and all nakshatras to be hidden, so also the universal form swallowed the form of the earth.'

This statement is also correct. During the day, the sun—because of its radiance—hides the moon and the nakshatras as though it has consumed them. Here, it is taken that 'srishti' means earth; in the infinite expanse of the universe, it, too, would be lost and unrecognizable, which is not surprising.

तयांही माजीं एकैके । सावियाची भयानकें ।
काळरात्रीची कटकें । उठावली जैसी कीं यें ।।१९९।।
मृत्युसिचि मुखे जाहली । हो कां जे भयाची दुर्गे पन्नासिली ।
कीं महाकुंडे उघडली । प्रळयानळाची ।।२००।।

In that were many faces—fearsome like an army that has revolted in the night of the apocalypse, or as if death itself had developed countless visages, as though terror had been given shape, or as if the cauldron holding the fire of dissolution had been flung open.

Arjuna, having obtained the opportunity to travel freely in space, would certainly have found what he saw frightening. Anyway, to travel in space one requires immense speed. We can imagine this while watching movies like 'Gravity' or 'Interstellar'. Outside the earth's atmosphere, moving in orbit at high velocity in a spacecraft where even tiny asteroids might strike you is terrifying. However, spacecraft in Earth orbit are not normally struck by asteroids; they are at risk mainly from micrometeoroids and space debris. To hang suspended in space and walk there is itself an unsettling experience. And having the chance to behold distant giant stars, enormous galaxies, or black holes in reality would be like gazing into the very mouths of death, because due to the extreme gravitational attraction of these massive objects, anything drawn inside cannot escape. They are, of course, forms and faces of death.

दिवी सूर्यसहस्त्रस्य भवेद युगपदुत्थिता ।
यदि भाः सदृशी सा स्याद भासस्तस्य महात्मनः ।।

How can we express Shri Krishna's body or his extreme radiance?

'It was as if it were a combination of 12 suns during the apocalypse, or even if 1,000 suns were to rise together, they would still pale before Shri Krishna's radiance.'

Now let us see how the sun would look during the apocalypse.

The apocalypse, when it comes, will be only on earth and nowhere else. When we speak of the apocalypse of the life of the sun, it refers to the time when its fuel is exhausted. During that time, it will increase in size and become more radiant; that is the condition being described. Its radiance would be unbearable from earth because its size would increase and it would be very close. Whenever the sun is mentioned in its normal radiance, it is always written as equal to 12 suns. Is it that the concept of the 12 suns refers to the 12 different types of rays that are emitted? If we look at the electromagnetic spectrum, then in that we have gamma rays, x-rays, ultraviolet, 7 colors of the visible spectrum, infra-red, and radio waves, which are in total 12 different types of rays. However, these are not distinct "rays emitted by the Sun" as separate solar types; they are classifications of electromagnetic wavelengths. Could this be what is meant by '12 suns'? But it is also true that even if 1,000 suns rise together, they will not be able to compete with the brilliance of Shri Krishna. This is because in the universe, there are not thousands but hundreds of billions of stars and countless galaxies.

लेलिह्यसे ग्रेसमान: समंताल्लोकां समग्रान वदनैर्ज्वलदभी
तेजो भिरापुर्य जगत समग्रँ भासस्तवोग्रा: प्रतपंति विष्णो ।।३०।।

The frightful form of the universal supreme figure of Vishnu has been described in this verse. Even after having swallowed a lot, his stomach is never full. This can be compared with the center of a galaxy or a black hole. However much it swallows, it is ready to swallow even more.

Big galaxies swallow smaller galaxies, and during such interactions smaller galaxies can lose or absorb individual stars rather than literally "swallowing" them, just as the Supreme Being is described as having innumerable mouths. Similarly, in this universe there are innumerable stars, galaxies, and black holes. These are actually the frightful mouths or faces of the Supreme Being.

It is astonishing to see such similar references.

We have just seen a passing reference. I wonder how many more similar astonishing facts might have been recorded by Sant Dnyaneshwar in the Dnyaneshwari. This is in itself a subject of research.

16

Astronomy In Dasbodh

Samarth Ramdas Swami completed writing the Dasbodh in Magha Shudha Navami (Ninth Day of the Magha month) in the 17th century, i.e., 1660. It is very important to give a detailed account of the composition of Dasbodh. Around the same time in Europe, Galileo, Edmond Halley, Cassini, Newton, and various other scientists were also working, and some original discoveries in astronomy were being made. In the same vein, it is important to note that the thirteenth chapter (Dashak) of Dasbodh, in its fourth composition, discusses the origins and progression of the universe in multiple Shlokas, which is worth mentioning. It is important to think about this in detail. Let us see how similar it is to modern science.

ब्रह्म घन आणि पोकळ। आकाशाहून विशाळ ।
निर्मळ आणि निश्चळ। निर्विकारी ।।१।।
ऐसेची असता कित्येक काळ । तेथे आरंभला भूगोळ ।
त्या भूगोळाचे मूळ। सावध ऐका ।।२।।

(Dashak 13 : Samas 4)

Brahma is tightly filled in all places, bigger than the sky, but since it is extremely minute it feels hollow and empty. After a lot of time has elapsed, the formation of Earth began; in other words, the solar system started to take shape.

Here Samarth was well aware of the fact that the universe is much larger than the region of sky visible from Earth. The sky is for the Earth just like a cover on it, but the universe is a hollow place which accommodates all that is—Earth, Sun, Moon, all planets, and stars.

According to modern science, the universe is mostly empty space, but matter exists within it. Brahma, which is bigger than the sky, has been described as pure, stationary, and unchanging. The universe is made up of "neutral particles." Was Samarth, by writing the word Nirvikar, wanting to convey "neutral"?

As per modern theory, there was a Big Bang, and this led to the slow formation of stars, star clusters, galaxies, etc. A long time after the Big Bang, our solar system was formed. How is this explained by Samarth?

वायोचा कातरा घसवटे । तेणे उष्णे वन्हि पेटे ।
सूर्यबिंब ते प्रगटे । तये ठाई ।।८।।

'Two gusts of wind coming from opposite directions rub against each other. Due to the friction, fire is generated. And hence the Sun forms.'

As per modern science, the solar system was formed from a giant molecular cloud, often called a solar nebula. An accretion disc formed from the nebula, and while rotating, from this emerged the Sun and the planets. The accretion disc, in short, is a disc of gas and dust revolving around a central point like a plate. At various places the particles revolving around the central point in spherical orbits came together. At the center, a huge body formed. This brought about extreme pressures on the particles due to gravitational attraction, the body heated up, and the process of nuclear fusion started. From the heat produced, a radiant Sun took birth. The Sun started shining as a star. Samarth has not said anything different.

वारा वाजतो सीतळ । तेथे निर्माण झाले जळ ।
ते जळ आळोन भूगोळ । निर्माण झाला ।।९ ।।

Along with the Sun, other planets and also Earth were formed. At the time of its formation, Earth was a molten ball of fire. Slowly it cooled and then began to freeze. The crust of the Earth solidified, and on that water formed. When land and water came together, living organisms were born.

नव्हे अनुमानाचे बोलणे । याचा बरा प्रत्यय घेणे ।
वेदशास्त्रे पुराणे । प्रत्यय घ्यावी ।।१७।।

Samarth says: This is not imaginary talk. This should be accepted only after actually studying the Vedas, Shastras, and Puranas.

It is important to remember that Samarth, who has written compositions like Ubharani Nirupan and Pralay Nirupan, himself says that we should not accept any principle without actual experience.

Description Of End Of 'Kalpa' In Dasbodh

Samarth, in his fourth composition, describes the end of the Kalpa. He says that the order in which the 'Origin' unfolded will be reversed in the sequence of its 'Collapse.' Can this be compared to modern astronomical theories of the 'Big Bang and Big Crunch'? The universe was formed from a Big Bang, and afterward stars, galaxies, and other cosmic structures slowly emerged. In the universe these galaxies contain innumerable stars, and most of them are moving away from each other, as scientists have discovered.

The future of the universe has been theorized by some scientists in recent observations: although these galaxies are receding from one another, it is possible that, at some distant point, they may return, converge, and create a condition akin to a 'Big Crunch.'

Samarth here also speaks of the future of the solar system and, therefore, of our own fate. At the very beginning of the fourth composition Samarth has said:

पृथ्वीस होईल अंत । भुतास मांडले कल्पांत ।।
ऐसा समाचार साध्यंत । शास्त्री निरोपिला ।।१।।
बारा कळी सूर्यमंडळा । किर्णापासूनी निघती ज्वाळा ।।
शतवरुषे भूगोळा । दहन होये ।।३।।

Samarth knew that the Earth would eventually come to an end—that a time will arrive when the Earth will simply cease to exist. From the sun's rays, fire will be emitted, and due to this the Earth will keep burning for 100 years.

We all know that when sunlight is dispersed, it separates into seven bands of color. At one end of the spectrum, beyond violet, lie ultraviolet rays, then X-rays and gamma rays; beyond red light are infrared and radio waves—each with different wavelengths invisible to the human eye. Combining the seven colors of visible light with the five invisible ranges—X-rays, gamma rays, ultraviolet, infrared, and radio waves—gives a total of twelve, exactly as described by Samarth. Should we consider this merely a coincidence in the phrase 'bara (12) kali surya mandala'?

After this he says, "Shata varushe bhugola dahan hoye"—and this is scientifically true. After the Sun exhausts its fuel, it will become a red giant. All stars follow a life sequence: as long as their fuel—hydrogen that fuses into helium—lasts, they continue producing heat. Once the fuel ends, the star's fate depends on its mass. After the fuel is depleted, the internal temperature begins to fall. Then, due to various chemical and physical processes, the star's outer layers expand. It becomes enormous. As this outer shell is cast off, its temperature drops, and it appears red. This stage is known as the Red Giant (Laal Rakshasi Tara). When the Sun enters this red giant phase, it will engulf Mercury, Venus, and the Earth itself. Earth will very likely be completely swallowed by the Sun. One can only imagine the planet's condition at that moment.

सिंधू नद्या एकवटल्या । नेणो नभींहुन रिचवल्या ।।
संधीच नाही धारा मिळाल्या । अखंड पाणी ।।१६।।
सप्त सिंधू आवर्णी गेले । आवर्ण वेडे मोकळे जाले ।।
जलरूप जालिया । खवळले प्रळये पावक ।।१८।।

Due to the increased size of the Sun, the temperature of the Earth will rise. First, all the ice at the poles will melt. The sea level will rise so dramatically that all seas, oceans, and rivers will merge, turning Earth into a single, unbroken sphere of water.

ब्रह्मांडाऐसा तप्त लोहो । शोषी जळाचा समुहो ।।
तैसे जळास जाले पाहो । अपूर्व मोठे ।।१९।।
तेणे आटोन गेले पाणी । असंभाव्य माजला वन्ही ।।
तया वन्हीस केली झडपणी । प्रळयवाते ।।२०।।

Due to the Sun's intense heat, this watery globe will evaporate. Under the apocalyptic fire, all water will vanish, leaving flames everywhere.

तेथे महाभूते खवळती । प्रळयेवात सुटती ।।
प्रळये पावक वाढती । चहुकडे ।।६।।
तेथे अक्रारुद्र खवळले । बारा सूर्य कडकडीले ।।
पावक मात्र येकवटले । प्रळयेकाळी ।।७।।
वायो विजांचे तडाखे । तेणे पृथ्वी अवघी तरके ।।
कठिणत्व अवघेचि फाके । चहुकडे ।।८।।

The five elements will boil and rise; the winds of destruction will howl; the fire of apocalypse will surge and consume all. During this final dissolution, fire will burn without pause. The Earth will tremble under relentless winds and lightning, and calamity will spread in all four directions. (Perhaps the lava from deep within the Earth will erupt and cover the planet?).

तेथे मेरुची कोण गणना । कोण सांभाळील कोणा ।।
चंद्र सूर्य तारांगणा । मूस जाली ।।९।।
पृथ्वीने विरी सांडिली । अवघी धगधगायेमान जाली ।।
ब्रह्मांडभटी जळोन गेली । येकसरा ।।१०।।

When the Moon, Sun, and Earth become one sphere ('Moos Jhali'), who will think of Mount Meru then? In the Sun's red giant phase, Mercury, Venus, and possibly Earth will be engulfed; it will follow Earth's fate because it is gravitationally bound to Earth. They will quite literally "become one vast sphere."

In the 10th verse Samarth says: "Earth becomes powder"—and indeed, what term could better capture the sense of 'wiri sandali'? Due to the overwhelming heat, the Earth begins to blaze. It burns in the furnace of the universe.

This description is indeed very much true. Inside the sun there burns an atomic furnace ablaze with nuclear fusion. After some time, the sun will also come to an end.

उदंड पोकळी थोडा वारा । तेणे वितळोन गेला सारा ।।
पंचभूतांचा पसारा । आटोपला ।।२२।।

In the earth's apocalyptic fire, all things will be reduced to ash. The wind of apocalypse will later make the ashes disappear. Earth's existence will end; it will merge into space. Space's emptiness is limitless. In this emptiness, this wind will vanish forever, ending the disarray caused by these five elements.

Samarth's description of earth's end is truly mind-boggling. The earth's future, as has been prophesied by modern science, is not at all different from this. A thousand bows to Samarth Ramadas Swami!

ऐसीं पंचमहाभूतें । पूर्वी होती अवेक्ते ।
पूढे जाली वेक्ते । सृष्टीरचनेसी ।।५७।।

(Dashaka 8, Samas 3-57)

This world is made up of five elements organized together. These elements are in the form of seeds. They form the Brahmand (The Universe) one after the other by combining in different quantities. One who understands this understands the formation of the Cosmos. The five elements are initially quiescent but later, to form the cosmos, they become manifest.

During the formation of the universe, all the elements (and energy) are quiescent; they all exist within a single point. Modern cosmology describes this as the extremely hot, dense state at the beginning of the Big Bang, literally "all elements in a single point." Later, after they are activated, they form stars and galaxies. They come into existence in actuality. Today, we know that when our solar system was formed, and Earth was formed out of it, it was a hot, molten sphere. There was nothing on it initially, and then slowly it cooled, and upon it formed soil, water, air, and atmosphere. All that had been obscured became visible and expressed.

With modern technology and the study of modern science, all this has been understood. All our knowledge is due to science. We have not actually seen or experienced it, but have studied it. How could Samarth write all these concepts so exactly when, in the olden times, there was no technology or modern instruments?

गगना सारिखे ब्रह्म पोकळ। उदंड उंच अंतराळ।
निर्गुण निर्मळ निश्चळ। सदोदित।।१।।

त्यास परमात्मा म्हणती। आणिक नामे नेणो किती।
परी ते जाणिजे आदिअंती। जैसे तैसे ।।२।।
विस्तीर्ण पसरलेला पैस। भोवता दाटला अवकाश।
भासचि नाही निराभास । जैसे तैसे ।।३।।
चहुकडे पाताळतळी। अंतचि नाही अंतराळी ।
कल्पांत काळी सर्व काळी। संचलेचि असे ।।४।।

(Dashak 10, Samas 10)

Brahma (Cosmic Consciousness) is like the sky. Just as air forms in the sky, similarly things form in the universe. As the sky is empty and stretches beyond all imagination, so is the Cosmic Consciousness. This Consciousness is always silent, stationary, and without disorder. This is Parmatma (God). No one knows by how many names it is called; irrespective of the number of names, it is from start to end the same. Imagine an infinite space spread all over. There is immense emptiness surrounding this space. From this infinite expanse, you will experience some portion of the universe. But there is a difference: space cannot be sensed. It feels empty to the mind, but you cannot experience it in any other way. That which cannot be sensed remains exactly as it is for all time. Spread in all four directions, below pataal (hell) and beyond antraal (space), is Parabrahma (formless, incorporeal, and the absolute source of all existence), which has no deformities. Not only at all times but even during the apocalypse (Kalpanta), it remains filled forever, exactly as before.

Until recently, we did not have much knowledge beyond that of Earth. After the rise of modern science and scientific instruments, we have slowly begun to understand Earth, the solar system, and what lies beyond. During Samarth's time, there was no modern knowledge. Still, he wrote firmly that space is a void and immeasurably vast.

नवखंड हे वसुंधरा। सप्तसागरांचा फेरा।
ब्रह्मांडाबाहेरील नीरा। कोण पाहे ॥१५॥
त्या निरामध्ये जीव असती। पाहो जाता असंख्याती।
त्या विशाल जीवांची स्थिती। कोण जाणे ॥१६॥
जेथे जीवन तेथे जीव। हा उत्पत्तीचा स्वभाव।
पाहाता याचा अभिप्राव। उदंड असे ॥१७॥

Earth is spread across nine continents and is surrounded by seven seas. But no one has ever seen the water that lies outside the earth in the vast universe. In that water, there are actually many organisms. No one knows the state of these vast beings. Where there is water, there is life. This is the essential feature of evolution. The mystery of this principle is profoundly deep.

In this Shloka, it is correct to say that Samarth has actually prophesied the future. There are nine continents on Earth, and surrounding them are seven seas, which have now been proven. But the real marvel comes later. Here Samarth says that "there is water outside space"—that is, beyond Earth and the solar system, there should also be water. Until recently, scientists were not ready to accept that there could be water anywhere other than on Earth.

Goldilocks Zone or Habitable Zone is the place where water can be found in the liquid state. If any planet is too close to its star, then all the water will evaporate, and if it is too far away, the water will freeze. The idea that planets in the Goldilocks Zone should host liquid water has been put forward with renewed confidence only very recently. In our solar system, Earth is the only planet whose surface stably supports liquid water today; because of this, scientists long believed that water in liquid form would not exist anywhere other than on Earth.

But now the Cassini spacecraft has seen plumes rising from geysers on Saturn's moon Enceladus. It has become clear that Enceladus has a subsurface sea. Gradually, we have discovered that there is water in many places in our solar system. It has been postulated that Jupiter's moon Europa also has a subsurface ocean. It is even believed that there is some water ice in the craters on Mercury (the planet closest to the Sun). It has also been speculated that there is water ice on the Moon. Evidence of water has been observed on planets outside our solar system as well. This knowledge is quite recent.

Can we not infer from this that Samarth had foreseen the future? In the future, humankind may find other planets with water and even vast living organisms. Who can say? Modern astronomy believes the saying that "where there is life there are organisms." They add that the reasoning behind this principle is profoundly deep. What more can be said? We offer many salutations to Samarth.

उष्ण प्रकाश तो सूर्याचा। शीतल प्रकाश तो चंद्राचा।
उष्णत्व नसता देह्याचा । घात होये ।।१०।।
याकारणे सुर्येविण। सहसा न चले कारण।
श्रोते तुम्ही विचक्षण। शोधुन पहा ।।११।।
हरिहर यांच्या अवतार मूर्ती । शिवशक्तीच्या अनंत वेक्ती ।
यापूर्वी होता गभस्ती। आतां हि आहे ।।१२।।
जितुके संसारासी आले। तितुके सूर्याखाले वर्तले।
अंती देहे त्यागून गेले। प्रभाकरा देखतां ।।१३।।
चंद्र ऐलीकडे जाला। क्षीरसागरा मधून काढिला।
चौदा रत्नांमध्ये आला। बंधू लक्ष्मीचा ।।१४।।
विश्वचक्षू हा भास्कर। ऐसे जाणती लहानथोर।
या कारणे दिवाकर। श्रेष्ठाहून श्रेष्ठ ।।१५।।
अपार नभमार्ग क्रमणे। ऐसेची प्रत्यही येणे जाणे।
या लोकोपकारा कारणे। आज्ञा समर्थाची ।।१६।।

(Dashak 16, Samas 2)

Sun's light is hot, while moonlight is cold. If the body does not have heat, it will not remain alive.

Hence, without the Sun nothing will move in this world. You all are critical readers. You can yourself derive the meaning of this.

"The entire history of the world has occurred under the eyes of the Sun. To bless human beings, every day the Sun rises in the sky. He is the ancestor of Sri Rama. We should bow before him and take his blessings. In this world, there are many forms of Harihara. Shiva and Shakti have also appeared in various forms, but before all of them was the Sun. He is still there as he was before."

"All that has come to life in visible form has arrived and departed under the Sun's gaze. The animals that enter this world all pass away under the Sun."

"The Moon formed later. After the churning of the Kshirsagar (ocean of milk), it arose. Among the fourteen gems that emerged, the Moon too appeared as Lakshmi's brother."

"The Sun is the eye of the world, and all—young and old—are aware of it. Hence the Sun is the greatest of the great."

"The Sun travels across vast distances in the sky. Every day it rises and sets along the same path. The Sun is the emblem of radiance, and there is nothing comparable to it."

Here Samarth talks about the importance and antiquity of the Sun. Samarth tells us, "In this world there have been different forms of Harihara, and Shivshakti appeared, but before them all was the Tejobhasker (Radiant Sun)." As per science, this is correct. "The Sun formed first, and the Earth formed later from the remaining material in the solar nebula, followed by humans, who began to think, and then Gods were conceived."

The mention that the Moon is quite recent is also scientifically accurate. The Earth was struck by a Mars-sized protoplanet (Theia), and debris from both bodies came together to form the Moon. When Earth was struck by that planet, the event would indeed have resembled the churning of the ocean of milk (Kshirsagar). Therefore, we may accept that the Moon was formed from the "stomach" of the Earth.

The statement that the Sun crosses vast distances in the sky each day is also true. Here it has been written that the Sun "walks" (in literal translation), a concept known even to earlier people, and thus Samarth also intuitively envisions a heliocentric model. That the Sun is at the center and the Earth moves around it took a long time for Western thinkers to accept, as we all know from history.

ऐका प्रळयाचें लक्षण । पिंडीं दोनी प्रळये जाण ।
येकनिद्रा येक मरण । देहांतकाळ ॥ १ ॥
देहाधारक तिनी मूर्ती । निद्रा जेव्हां संपादिती ।
तो निद्राप्रळय श्रोतीं । ब्रह्मांडींचा जाणावा ॥ २ ॥
तिनी मूर्तीस होईल अंत । ब्रह्मांडास मांडेल कल्पांत ।
तेव्हां जाणावा नेमस्त । ब्रह्मप्रळये जाला ॥ ३ ॥
दोनी पिंडीं दोनी ब्रह्मांडीं । च्यारी प्रळय नवखंडीं ।
पांचवा प्रळय उदंडी । जाणिजे विवेकाचा ॥ ४ ॥
ऐसे हे पांचहि प्रळये । सांगितले येथान्वयें ।
आतां हें अनुभवास ये । ऐसें करूं ॥ ५ ॥
निद्रा जेव्हां संचरे । तेव्हां जागृतीव्यापार सरे ।
सुषुप्ति अथवा स्वप्न भरे । अकस्मात आअंगीं ॥ ६ ॥
या नांव निद्राप्रळये । जागृतीचा होये क्षये ।
आतां ऐका देहांतसमये । म्हणिजे मृत्युप्रळये ॥ ७ ॥
देहीं रोग बळावती । अथवा कठीण प्रसंग पडती ।
तेणें पंचप्राण जाती । व्यापार सांडुनी ॥ ८ ॥
तिकडे गेला मनपवनु । इकडे राहीली नुस्ती तनु ।
दुसरा प्रळयो अनुमानु । असेचिना ॥ ९ ॥
तिसरा ब्रह्मा निजेला । तों हा मृत्यलोक गोळा जाला ।
अवघा व्यापार खुंटला । प्राणीमात्रांचा ॥ १० ॥
तेव्हां प्राणीयांचे सुक्ष्मांश । वायोचक्रीं करिती वास ।
कित्येक काल जातां ब्रह्मयास । जागृती घडे ॥ ११ ॥
पुन्हा मागुती सृष्टि रची । विसंचिले जीव मागुतें संची ।
सीमा होतां आयुष्याची । ब्रह्मप्रळय मांडे ॥ १२ ॥

शत वरुषें मेघ जाती । तेणें प्राणी मृत्यु पावती ।
असंभाव्य तर्के क्षिती । मर्यादेवेगळी ॥ १३ ॥
सूर्य तपे बाराकळी । तेणें पृथ्वीची होय होळी ।
अग्नी पावतां पाताळीं । शेष विष वमी ॥ १४ ॥
आकाशीं सूर्याच्या ज्वाळा । पाताळीं शेष विष वमी गरळा ।
दोहिकडून जळतां भूगोळा । उरी कैंची ॥ १५ ॥
सूर्यास खडतरता चढे । हलकालोळ चहुंकडे ।
कोंसळती मेरूचे कडे । घडघडायमान ॥ १६ ॥
अमरावती सत्यलोक । वैकुंठ कैळासादिक ।
याहिवेगळे नाना लोक । भस्मोन जाती ॥ १७ ॥
मेरु अवघाचि घसरे । तेथील महीमाच वोसरे ।
देवसमुदाव वावरे । वायोचक्रीं ॥ १८ ॥
भस्म जालिया धरत्री । प्रजन्य पडें शुंडाधारीं ।
मही विरे जळांतरीं । निमिष्यमात्रें ॥ १९ ॥
पुढें नुस्ते उरेल जळ । तयास शोषील अनळ ।
पुढें एकवटती ज्वाळ । मर्यादेवेगळे ॥ २० ॥
समुद्रींचा वडवानळ । शिवनेत्रींचा नेत्रानळ ।
सप्तकंचुकींचा आवर्णानळ । सूर्य आणी विद्युल्यता ॥ २१ ॥
ऐसे ज्वाळ एकवटती । तेणें देव देह सोडिती ।
पूर्वरूपें मिळोन जाती । प्रभंजनीं ॥ २२ ॥
तो वारा झडपी वैश्वानरा । वन्ही विझेल येकसरा ।
वायो धावें सैरावैरा । परब्रह्मीं ॥ २३ ॥
धूम्र वितुळे आकाशीं । तैसे होईल समीरासी ।
वहुतां मधें थोडियासी । नाश बोलिला ॥ २४ ॥
वायो वितुळतांच जाण । सूक्ष्म भूतें आणी त्रिगुण ।
ईश्वर सांडी अधिष्ठान । निर्विकल्पीं ॥ २५ ॥
तेथें जाणिव राहिली । आणी जगज्जोती निमाली ।
शुद्ध सारांश उरली । स्वरूपस्थिती ॥ २६ ॥
जितुकीं काहीं नामाभिधानें । तये प्रकृतीचेनि गुणें ।
प्रकृती नस्तां बोलणें । कैसें बोलावें ॥ २७ ॥
प्रकृती अस्तां विवेक कीजे । त्यास विवेकप्रळये बोलिजे ।
पांचहि प्रळय वोजें । तुज निरोपिलें ॥ २८ ॥

इति श्रीदासबोधे गुरुशिष्यसंवादे
पंचप्रळयनिरूपणनाम समास पांचवा ॥

(Dashak 10, Samas 5)

Here Samarth has described the concept of 'panch pralaya', the five types of apocalypses. After the world is formed, it unfolds according to its own laws. Since the progress of time is cyclic, after the world is formed, it also ends after some time. This concept of the end of the world is clearly mentioned at the very beginning in the Taittiriya Upanishad. When the day of Lord Brahma starts, the world is formed, and when his day ends and night begins, the world comes to an end. The next time after the night has passed and the day begins, the world is recreated. In this, the same civilization forms again. Lord Brahma's day is called 'Kalpa' or 'Sarga', and his night is called 'Pralaya'.

As per our calendar system, a Kalpa lasts for 4.32 billion years. During all these years the universe continues, and once Lord Brahma's day ends, the night starts. When the Pralaya begins, the order in which the universe dissolves is exactly the reverse of the order in which it was formed.

Here it is essential to remember the most important principle in spiritual sciences: that 'Pinda' means the 'Microcosm' and 'Brahmand' means the 'Macrocosm', and both are formed in exactly the same manner. The human being is the symbol of the universe's formation. In a human being, all the principles of the universe are seen.

Samarth speaks of five apocalypses. Of these, two belong to 'Pinda' (Microcosm) and two to 'Brahmand' (Macrocosm), with one of 'Vivek' (Consciousness). The apocalypses of Pinda and Brahmand occur naturally, but the apocalypse of Vivek must be achieved by one's own effort.

The two apocalypses of the Pinda are 'Deep Sleep' and 'Death'.

Even the universe has two apocalypses. When Brahma, Vishnu, and Mahesh—the three Gods who have taken form—go to sleep, the Brahmand, that is, the Universe, goes into an apocalypse. This is the time when Lord Brahma's day ends and night begins. When all three Gods who have assumed form—that is, Brahma, Vishnu, and Mahesh—end, then the 'Kalpanta' of the universe starts. At this point we may understand that 'Brahma Pralaya' has begun.

The third Pralaya starts when Lord Brahma goes to sleep. During that time, all people on earth die and all life comes to an end. All creatures then remain as small particles in the universe. In this state a great span of time passes. Then, after the night of Lord Brahma ends, the day starts. He awakens, and after rising, he again creates the world. The lives of all the creatures scattered everywhere are again amassed by Lord Brahma. This is the Universe's Sleep Apocalypse.

When the life of Lord Brahma ends, Brahma Pralaya starts. Now consider the detailed description of Brahma Pralaya. For 100 years there are no clouds, and the sun shines with 12 different kinds of rays and creates heat. Due to this, the earth burns. As the sun's heat increases, there is chaos everywhere. All fires come together in one place. During that time, even the Gods have to leave their bodies. After this, all the Gods return to their original state and merge with the Primordial Wind.

Now begins the rule of the Primordial Wind. The wind starts blowing vigorously, envelops the fire, and extinguishes it. After this, within the Parabrahma (The Supreme Creator beyond conceptualization), a tremendous primordial wind begins to blow; but just as in a time of plenty a small amount of smoke vanishes into the sky, similarly the primordial wind inside Parabrahma disappears entirely. The wind, in this way, vanishes.

These five subtle elements, the three Gunas (qualities of nature—Satva: calmness; Rajas: ambition, aggression; Tamas: laziness, confusion), and Ishwar (Supreme Being or God) discard their distinctions and dissolve into the Parabrahma. The memory of their state of Ultimate Purity remains, but knowledge and wisdom are lost. In this manner, after the ultimate apocalypse, only the original state remains. The resolution that God once made to create the universe is the true resolution. It is a resolution of wisdom, consciousness, and endowed power, which is called 'Jagat Jyoti'. During the time of apocalypse, the universe disappears into itself, and God merges with the Brahma Roop—the universal form.

This is the description in the Dasbodh.

During the time of the writing of the Taittiriya Upanishad, to imagine the end of the universe was an extraordinary achievement. Without understanding the vastness and expanse of the universe, it is difficult to conceptualize its end. In ancient times there was such profound understanding, and what existed then is astonishing in itself.

"Pinda and Brahmand are similar" is a profound thought, and the ability of the ancient sages to understand this is remarkable.

The idea that time progresses cyclically is another important discovery. To arrive at this concept, modern scientists are still progressing gradually. The theory of 'Big Bang' to 'Big Crunch' has not yet been completely proven, but is it not essentially the same idea—that time is cyclical? The theory of the Big Bang to Big Crunch states that the universe forms from a single point called a singularity, expands to a certain limit, stops, and then begins to contract in the opposite direction until it returns to a single point. And then again—Hari Om!

Reading and reflecting on all this is enough to make one dizzy. It is sad to see how much of our ancient knowledge was forgotten, lost, and burnt. A great deal of damage has been done which can never be recovered. Yet, even with all that is lost, the remaining knowledge is immense, and to understand even this remainder, we will have to study deeply.

Details of References

- Valmiki Ramayana, Researched edition, Bhandarkar Oriental
- Research Institute (BORI) Pune
- Mahabharata, Researched edition, Bhandarkar Oriental
- Research Institute (BORI) Pune
- Rigveda Samhita, Yajurveda Samhita, Maitrayani Aaranyak
- Upanishad
- Surya Siddhanta, Panchsheel Siddhanta, Mahabhaskariya
- Siddhanta, Brahmasfut Siddhanta
- Siddhanta Shiromani, Yogavasishtha, Parashar Tantra
- BruhatSamhita :Varahmir:,Aryabhattiya:,Aryabhatt
- Shrimad Bhagwat Purana, Harivansh
- Mahabharattatparyanirnay : Acharya Madhva
- Baal - Ramayana : Rajshekhar
- Dnyaneshwari : Sant Dnyaneshwar, Dasbodha : Samarth
- Ramdas
- P. V. Vartak : Swayambhu, Pune
- P.V. Vartak :Vaastava Ramayana

- Basham A.L. : The Wonder that was India
- Dikshit, S.B. :History of Indian Astronomy
- Oak, Nilesh Nilkanth : When did the Mahabharata war Happen? : The Mystery of Arundhatti
- Oak, Nilesh Nilkanth : The Historic Rama
- Oak, Nilesh Nilkanth : Bhishma Nirwana
- P. Clift, A. Carter, and others : The paper U-Pb Zircon dating
- evidence for a Pleistocene Sarasvati River and Capture of
- Yamuna River, Geology, 2012
- P. Priyadarshi Dr : India's Contribution to the West
- Schroeder, Leopold von Schroeder, Pythagoras und die
- Inder(Pythagoras and the Indians)
- Vaidya, C.V. Bharatacharya: Epic India: India as described in
- the Mahabharata and the Ramayana
- Vaidya, C.V. Bharatacharya : The Mahabharata : A criticism
- A few Papers and lectures at different seminars

About The Author

Nilesh Nilkanth Oak
Email : drno5561bce@gmail.com

Nilesh Nilkanth Oak holds an M.S. in Chemical Engineering and an Executive MBA. He is a renowned researcher, author, TEDx speaker, and a distinguished alumnus of UDCT-ICT, widely popular for his eloquent oratory skills. His work focuses on bringing awareness to the extraordinary knowledge, technology, and antiquity of Indian culture. He is known for presenting arguments rooted in sharp intellect, objective evidence, and rigorous scientific and logical reasoning.

With a deep study of various global cultures and diverse subjects, he serves as a researcher and adjunct professor at the Institute of Advanced Sciences in Dartmouth, Massachusetts. He has authored three revolutionary books based on his fundamental research: When did the Mahabharata War Happen?, The Historic Rama, and Bhishma Nirvana. These works have been, and continue to be, translated into multiple languages. He travels extensively across the globe, engaging with students at prestigious universities and colleges, as well as interacting with scholars and diverse audiences. His ground-breaking research has served as an inspiration for various novels, documentaries, and films.

About The Author

Rupa Bhaty
Email : rupabhaty16@gmail.com

Rupa Bhati is an accomplished architect and assistant professor affiliated with the Institute of Advanced Sciences in Dartmouth, Massachusetts. She has dedicated herself to the deep study of Surya Siddhanta, Vedanga Jyotisha, and other ancient Indian Sanskrit texts. Her academic contributions include extensive research papers published in the field of archaeology, alongside a focused study spanning 30 years on the relationship between the Rigveda and human evolution.

In addition to her academic pursuits, she is deeply involved in architectural design and archaeological projects. A passionate environmentalist, she has been actively engaged in conservation efforts, including the planting of 2,000 trees in the Kutch region. During her tenure as the Chief Architect for the Gandhidham Development Authority, she pioneered a green initiative by making it mandatory to plant at least one tree in every household.

Beyond her professional and environmental work, Rupa is a talented canvas painter who has held exhibitions in Mumbai, Delhi, and Gandhidham, earning praise from both dignitaries and critics. Her resilience was notably displayed following the devastating earthquake of 2000, where she worked tirelessly with the City Development Committee and various NGOs to rebuild Gandhidham. Her efforts were recognized internationally by the Association for Women in Architecture (AWA), USA. She is also a distinguished member of the International Federation for Housing and Planning (IFHP) based in The Hague.

About The Author

Leena Damle
Email : lee.dams@gmail.com

Leena Damle holds an M.Sc. in Nuclear Physics and has completed a comprehensive three-year course in German language studies. She is a prolific science communicator and author, with notable published works including Katharupi Khagolshastra (Astronomy through Stories), Antarikshachya Antarangat (Into the Depths of Space), and Nakshatrakatha (Tales of the Constellations). She also translated the renowned scientist Michio Kaku's book Physics of the Impossible into Marathi under the title Ashakya Bhautiki. Her academic contributions include drafting various astronomy-related entries for the Marathi Vishwakosh (Marathi Encyclopedia).

A well-known figure in science journalism, she has authored numerous article series for leading dailies and monthly magazines and has contributed extensively to science-based programming for Akashwani (All India Radio). Her literary excellence has been recognized with the prestigious Raja Kelkar Award (2006) from the State Government for Katharupi Khagolshastra. Additionally, her collection of stories for children, Nakshatrakatha, was selected by the state government for distribution across all schools. In 2019, she was honored by the Maharashtra Sahitya Parishad, Pune, with the Late Go. Ra. Paranjape Special Author Award (sponsored by Shri M. N. Gogate) for her outstanding contributions to science writing.

On the international and national stage, Leena has represented India as a subject matter expert. In 2009, she represented the country at the "Stars of Asia" workshop in Tokyo, Japan, which focused on the astronomical myths and legends of Asian nations. She was also invited to conduct a specialized session on the intersection of astronomy and mythology at the India International Science Festival.

Books by Neelesh Oak

Published by Sakal Publications

- प्राचीन भारतीय खगोलविज्ञान
- Sugriva's Atlas

Also by Sakal Publications

Fiction

- Genesis of Kalki

Indian History

- Flag of Ananta
- Śītalā

Spiritual

- Uddhava Gita
- Discovery of Anugita
- Ramcharitmanas
- Awakening Your Bliss
- The untold secrets of life : Purushasuktam
- Simple Spirituality : Recalling Kabir
- Zen : Practicle Principles for Balanced Living

www.ingramcontent.com/pod-product-compliance
Lightning Source LLC
Chambersburg PA
CBHW031350230926
51848CB00002B/470
9788169355476